DEVELOPMENTS IN STEM CELL RESEARCH

DEVELOPMENTS IN STEM CELL RESEARCH

PRASAD S. KOKA
EDITOR

Nova Biomedical Books
New York

Library of Congress Cataloging-in-Publication Data

ISBN: 978-1-60456-341-2

Published by Nova Science Publishers, Inc. ✦ New York

Contents

Preface

The two broad categories of mammalian stem cells exist: embryonic stem cells, derived from blastocysts, and adult stem cells, which are found in adult tissues. In a developing embryo, stem cells are able to differentiate into all of the specialised embryonic tissues. In adult organisms, stem cells and progenitor cells act as a repair system for the body, replenishing specialised cells. As stem cells can be readily grown and transformed into specialised tissues such as muscles or nerves through cell culture, their use in medical therapies has been proposed. In particular, embryonic cell lines, autologous embryonic stem cells generated therapeutic cloning, and highly plastic adult stem cells from the umbilical cord blood or bone marrow are touted as promising candidates. Among the many applications of stem cell research are nervous system diseases, diabetes, heart disease, autoimmune diseases as well as Parkinson's disease, end-stage kidney disease, liver failure, cancer, spinal cord injury, multiple sclerosis, and Alzheimer's disease. Stem cells are self-renewing, unspecialized cells that can give rise to multiple types all of specialized cells of the body. Stem cell research also involves complex ethical and legal considerations since they involve adult, fetal tissue and embryonic sources. This new book presents the latest research in the field from around the globe.

Chapter 1 - Human embryonic stem cells (hESCs) may offer an unlimited supply of cells that can be directed to differentiate into all cell types within the body and used in regenerative medicine for tissue and cell replacement therapies. Previous work has shown that exposing hESCs to exogenous factors such as dexamethasone, ascorbic acid and β-glycerophosphate can induce osteogenesis. The specific factors that induce osteogenic differentiation of hESCs have not been identified yet, however, it is possible that differentiated human bone marrow stromal cells (hMBSCs) may secrete factors within the local microenvironment that promote osteogenesis. Here we report that the lineage progression of hESCs to osteoblasts is achieved in the presence of soluble signaling factors derived from differentiated hBMSCs. For 28 days, hESCs were grown in a transwell co-culture system with hBMSCs that had been previously differentiated in growth medium containing defined osteogenic supplements for 7-24 days. As a control, hESCs were co-cultured with undifferentiated hBMSCs and alone. Von Kossa and Alizarin Red staining as well as immunohistochemistry confirmed that the hESCs co-cultured with differentiated hBMSCs formed mineralized bone nodules and secreted extracellular matrix protein osteocalcin (OCN). Quantitative Alizarin Red assays showed increased mineralization as compared to the control with undifferentiated hBMSCs. RT-PCR revealed the loss of pluripotent hESC markers with the concomitant gain of osteoblastic markers such as collagen type I, runx2, and osterix. We demonstrate that osteogenic growth factors derived from differentiated hBMSCs within the local microenvironment may help to promote hESC osteogenic differentiation.

Chapter 2 - A high expression of hyaluronic acid (HA) in precartilage condensation at the early stage of endochondral ossification indicates its importance in bone and cartilage development. However, it remains unclear whether HA is associated with bone development by promoting the differentiation from mesenchymal stem cells (MSCs). In this study, we examined the effects of HA on the differentiation of MSCs into osteoblasts and chondrocytes in a three-dimensional collagen gel culture system.

Human MSCs were cultured in HA-collagen hybrid gel (0.15% collagen, 0.5 mg/ml HA) and collagen mono gel (0.15% collagen). The cells were treated by either osteogenic differentiation medium (ODM) or chondrogenic differentiation medium (CDM). The HA-collagen gel maintained in ODM exhibited greater alkaline phosphatase activity, calcium deposition, and gene expression of bone markers than the collagen mono gel. The HA-collagen gel maintained in CDM had greater glycosaminoglycan deposition and gene expression of cartilage markers than the collagen mono gel.

These findings indicate that the addition of HA to collagen gel would be highly recommended for promoting the differentiation of MSC to osteoblasts and chondrocytes in three-dimensional condition.

Chapter 3 - Embryonic stem cells (ESC) unlike normal somatic cells and their adult counterparts can proliferate indefinitely in culture without compromising on their pluripotency. While maintaining their self renewing capacity during expansion, they do not appear to undergo the process of aging. Evidently, this phenomenon of escaping senescence in hESC is possible because of its high telomerase activity, cell cycle regulation, DNA repair mechanism, as well as the lack of cytogenetic, genomic, mitochondrial and epigenomic changes. However, this capacity in hESC is lost with the onset of differentiation into fully matured somatic cells. Thus, hESCs hold tremendous promise in cell replacement therapy for various degenerative disorders like Parkinson's and Alzheimer's disease, cardiac disorders, type-I diabetes and many more. Although the process of self renewal and senescence in hESC is known to be regulated by several signaling pathways, either independently or cohesively, the molecular mechanisms enabling this process are not clearly delineated. This article provides a summarized overview of the interplaying mechanisms known to govern self renewal in hESC, with special emphasis on the role of cell cycle regulation.

Chapter 4 - Hematopoietic stem cells (HSCs) are characterized by two distinct abilities, that is, self-renewal ability and multipotency. To keep homeostasis of hematopoiesis and protect exhaustion of HSCs throughout the life, most of HSCs are kept quiescent and only a limited number of HSCs enter cell cycle to supply mature blood cells. Cell cycle state of HSCs is crucially regulated by external factors such as cytokines, adhesion molecules, Notch ligands, and Wnt signals in the bone marrow (BM) microenvironment, so called hematopoietic niche. In addition, intrinsic molecules expressed in HSCs such as transcription factors and cell cycle regulatory molecules also control their growth and differentiation. To utilize HSCs more efficiently and to develop new therapeutic strategies for various diseases, it is of particular interest to expand HSCs *ex vivo*. At present, three clinical studies, in which cord blood HSCs were *ex vivo* expanded by cytokines and transplanted into patients with hematologic malignancies, have been performed. However, the expanded HSCs did not shorten the recovery of hematopoiesis. So, further novel strategies to expand HSCs more efficiently and to fasten hematopoiesis from HSCs are required by modifying the function of the molecules that regulate self-renewal of HSCs.

Chapter 5 - Purified bone marrow-derived mesenchymal stem cells were autotransplanted to a patient with acute myocardial infarction. The employment of the sub-population of bone marrow-derived mononuclear cells was intended to clarify some disputable outcomes in heterogeneous mononuclear cell therapy. The improvement of myocardial perfusion and cardiac function was observed after delivery of mesenchymal stem cells through a combined procedure of primary intracoronary infusion and secondary intravenous infusion. This procedure is expected to enhance the engraftment efficacy of transplanted cells at infarcted myocardium.

Chapter 6 - Fetal liver-derived adherent fibroblast-like stromal cells (FLSCs) were transfected with retrovirus, carrying the human telomerase catalytic subunit (*hTERT*) gene to establish an immortalized mesenchymal stem-like cell line, designated as FLMSL-hTERT. This cell line had the potential to differentiate into osteocytes and adipocytes. Their immunophenotype was similar with that of mesenchymal stem cells (MSCs) derived from bone marrow (BM), umbilical cord blood, skin, and adipose tissue. The cell line expressed mRNAs of SCF, Wnt5A, FL, KIAA1867, TGF-β, Delta-like, VEGF, SDF-1, PLGF and Jagged-1 that are known to promote hematopoiesis. The FLMSL-hTERT cells could dramatically support cord blood (CB) HSCs/HPCs expansion *ex vivo*, especially to maintain HSC properties *in vitro* at least for 8 wk; moreover, they appeared to be superior to primary human fetal liver stromal cells (FLSCs) in supporting *in vitro* hematopoiesis, because they were more potent than primary FLSCs in supporting the *ex vivo* growth of HSC/HPCs during long-term culture. The FLMSL-hTERT cells have been maintained *in vitro* more than 150 population doublings (PDs) with the unchanged phenotypes. So this cell line may be of value for *ex vivo* expansion of HSCs/HPCs and for analyzing the human fetal liver hematopoietic microenvironments.

Chapter 7 - At this time point renal injury, ultimately resulting in renal failure, can not be therapeutically cured by endothelial, glomerular or tubular regeneration. Over the last five years, several reports appeared in which the replacement of damaged renal cells by bone marrow-derived cells (BMDC) was shown, thereby suggesting a therapeutic role for BMDC in renal regeneration. For correct interpretation of the function of these cells in renal repair, *in vivo* tracking of BMDC is crucial. Since various tracking methods with variable experimental outcomes have been reported, we will provide an overview of these methods and discuss their advantages and drawbacks for experimental renal disease models.

Chapter 8 - MicroRNAs (miRNAs) are endogenous ~22 nucleotide non-coding RNAs. They play important regulatory roles in plants and in animals by pairing to messenger RNAs of target genes, specifying mRNA cleavage or repression of protein synthesis. Recent evidences indicate that they exhibit important regulatory roles in development timing, cell proliferation, cell survival and apoptosis. miRNAs regulate normal stem cell development both in mammalian and non-mammalian systems. For instance, they regulate stem cell self-renewal and differentiation, and have gained increasing attention in stem cell biology. Here we review recent progression of miRNAs in stem cells research, and summarize current understanding of miRNA in their expression in stem cells and their cellular biological role in stem cells.

Chapter 9 - Aplastic Anaemia is correlated with Stem Cell disorder, the detail mechanism of which still awaits extensive evaluation. Apart from "Inherited Aplastic Anaemia", the

"Acquired Aplastic Anaemia", has also been considered to involve stem cell deficiency or disorder or more currently considered disorganized "stem cell niche". The manifestations in such types of Aplastic Anaemia are varied and supposed to impair the immunological functions as well. Causes for Acquired Aplastic Anaemia can also be multifactorial that may concern the stem cell itself or its microenvironment or the stem cell "niche", which is, however, in controversy. Studies conducted in our laboratory revealed that the stromal microenvironment is significantly affected in Acquired Aplastic Anaemia, especially in experimental animals and farmers exposed to insecticides/ pesticides of organophosphate and organochloride origin. Supplementation of healthy microenvironmental components including stromal cells and cord blood plasma factor (CBPF) have been found to rectify the stem cell deficiency under the event. It is suggested that Acquired Aplastic Anaemia can be recovered reversibly to health by supplementation of microenvironmental factors together with immune-reconstitution. This review attempts to re-establish the involvement of niche system and the role of microenvironment /microenvironmental factors as therapeutic adjunct to the patient-concerned.

Chapter 10 - HIV infection and its outcome is complex because there is great heterogeneity not only in clinical presentation, incomplete clinical information of markers of immunodeficiency and in measurements of viral loads. Also, there many gene variants that control not only viral replication but immune responses to the virus; it has been difficult to study the role of the many AIDS restricting genes (ARGs) because their influence vary depending on the ethnicity of the populations studies and because the cost to follow infected individuals for many years. Nevertheless, at least genes of the major histocompatibility locus (MHC) such as HLA alleles have been informative to classify infected individuals following HIV infection; progression to AIDS and long-term-non-progressors (LTNP). For example, progressors could be defined as up to 5 years, up to 11 years or as we describe in this report up to 15 years from infection, and LTNP could be individuals with normal CD4+ T cell counts for more than 15 years with or without high viral loads. In this review, we emphasized that in the studies of ARGs the HLA alleles are important in LTNP; HLA-B alleles influencing the advantage to pathogens to produce immune defense mediated by CD8+ T cells (cognate immunity). Our main point we make in this report is that contrary to recent reports claiming that this dominant effect was unlikely due to differences in NK activation through ligands such as HLA-Bw4 motif, we believe that cognate immunity as well as innate inmunity conferred by NK cells are involved. The main problem is that HLA-Bw4 alleles can be classified according the aminoacid in position 80. Isoleucine determines LTNP, which is a ligand for 3DS1. Such alleles did not include HLA-B*44, B*13 and B*27 which have threonine at that position. The authors have not considered the fact that in addition to the NK immunoglobulin receptors, NK receptors can be of the lectin like such as NKG2A/HLA-E to influence the HIV infection outcome. HLA-Bw4 as well as HLA-Bw6 alleles can be classified into those with threonine or methionine in the second position of their leader peptides. These leader peptides are ligands for NKG2A in which methionine influences the inhibitory role of NKG2A for killing infected targets. Functional studies have not been done as well as studies of these receptors in infected individuals. However, analyses of the leader peptides of HLA-B alleles in published reports, suggested that threonine in the second position can explain the importance of HLA-B*57, B*13, B*44 as well as certain Bw6 alleles in LNTP. In addition, we analyzed the San Francisco database that was reported and found that the association of HLA-B alleles with LNTP or with progressors can be due to the presence of threonine or methionine in their second position. Therefore, studies of outcome of HIV infection should include not only mechanisms of cognate immunity mediated by peptides and CD8+ T cells but also, NK receptors of two types, NKG2A as well as 3DS1. We propose that the SCID mouse should be used to understand mechanisms mediated by many of the ARGs especially the importance of thymus derived cells as well as NK receptor interactions with their ligands in this experimental animal transplanted with human stem cells, thymus or NK cells obtained from individuals of known HLA genotypes.

In: Developments in Stem Cell Research
Editor: Prasad S. Koka

ISBN: 978-1-60456-341-2
© 2008 Nova Science Publishers, Inc.

Chapter 1

HUMAN EMBRYONIC STEM CELLS UNDERGO OSTEOGENIC DIFFERENTIATION IN HUMAN BONE MARROW STROMAL CELL MICROENVIRONMENTS

Wilbur Tong[#,1], Shelley E. Brown[#,1,2] and Paul H. Krebsbach[,1,2]*

[1]Biologic and Materials Sciences, University of Michigan School of Dentistry
Ann Arbor, MI 48109
[2]Biomedical Engineering, University of Michigan College of Engineering
Ann Arbor, MI 48109

ABSTRACT

Human embryonic stem cells (hESCs) may offer an unlimited supply of cells that can be directed to differentiate into all cell types within the body and used in regenerative medicine for tissue and cell replacement therapies. Previous work has shown that exposing hESCs to exogenous factors such as dexamethasone, ascorbic acid and β-glycerophosphate can induce osteogenesis. The specific factors that induce osteogenic differentiation of hESCs have not been identified yet, however, it is possible that differentiated human bone marrow stromal cells (hMBSCs) may secrete factors within the local microenvironment that promote osteogenesis. Here we report that the lineage progression of hESCs to osteoblasts is achieved in the presence of soluble signaling factors derived from differentiated hBMSCs. For 28 days, hESCs were grown in a transwell co-culture system with hBMSCs that had been previously differentiated in growth medium containing defined osteogenic supplements for 7-24 days. As a control, hESCs were co-cultured with undifferentiated hBMSCs and alone. Von Kossa and Alizarin Red staining as well as immunohistochemistry confirmed that the hESCs co-

[#] These authors contributed equally to this work.
[*] **Corresponding author:** Paul H Krebsbach, DDS, PhD. Department of Biologic and Materials Sciences, University of Michigan, School of Dentistry, 1011 N. University Ave K1030. Ann Arbor, MI 48109-1078 USA. Tel (734) 936-2600. Fax (734) 763-3453. paulk@umich.edu

cultured with differentiated hBMSCs formed mineralized bone nodules and secreted extracellular matrix protein osteocalcin (OCN). Quantitative Alizarin Red assays showed increased mineralization as compared to the control with undifferentiated hBMSCs. RT-PCR revealed the loss of pluripotent hESC markers with the concomitant gain of osteoblastic markers such as collagen type I, runx2, and osterix. We demonstrate that osteogenic growth factors derived from differentiated hBMSCs within the local microenvironment may help to promote hESC osteogenic differentiation.

Keywords: human embryonic stem cells, bone marrow stromal cells, osteoblasts, stem cell-microenvironment

INTRODUCTION

Human embryonic stem cells (hESCs) present a potentially unlimited supply of cells that may be directed to differentiate into all cell types within the body and used in regenerative medicine for tissue and cell replacement therapies. An area of particular interest is stem cell transplantation for bone tissue regeneration where hESCs may be used to repair skeletal defects. One of the major gaps in the knowledge regarding hESCs is the lack of understanding of the growth factors and three-dimensional signals that control differentiation. Current techniques used for bone tissue repair employ the use of auto- and allografting methods, however, these methods have inherent limitations that restrict their universal application. The limitations of these reparative strategies suggest that an alternative approach is required, and hESCs may provide a repository of cells for such an approach. Previous work has shown that exposing hESCs to exogenous factors such as dexamethasone, ascorbic acid and β-glycerophosphate can induce osteogenesis *in vitro* [1-3]. However, the specific factors that regulate and influence the commitment of hESCs along the osteoblast lineage have not yet been identified. It is possible that soluble factors secreted by human bone marrow stromal cells (hBMSCs) may provide the necessary signaling molecules to direct osteogenic differentiation of hESCs.

When bone marrow is cultured *in vitro*, adherent non-hematopoietic cells proliferate and exhibit characteristics of bone marrow stroma *in vivo*. Within this diverse population of hBMSCs there exist early progenitor mesenchymal stem cells that are capable of self-renewal and have multi-lineage differentiation potential into cell types such as osteoblasts, chondrocytes, and adipocytes. In the presence of dexamethasone, ascorbic acid and β-glycerophosphate, it has been demonstrated that hBMSCs can be differentiated readily into mineralizing osteoblasts both *in vitro* and *in vivo* [4-7]. The *in vitro* equivalent of bone formation is characterized by the formation of mineralized nodules, increased alkaline phosphatase activity, and up regulation of osteoblastic genes such as runx2, osteocalcin, bone sialoprotein, and collagen type I [5, 8-10]. The use of this well-defined *in vitro* model allows the control of the differentiation state of hBMSCs at varying time points within their lineage progression towards functional osteoblasts. Subsequently, soluble factors derived from hBMSCs may be controlled, thus enabling the establishment of a co-culture system that stimulates hESC differentiation.

Therefore, the goal of this study was to determine if the lineage progression of hESCs toward osteoblasts could be directed by soluble signaling factors derived from differentiated hBMSCs. Because the differentiation of hBMSCs *in vitro* is so well characterized, the ability to manipulate and control hBMSCs along with the osteogenic factors derived from them will be integral to controlling the osteoblastic differentiation of hESCs. In this study, we demonstrate that osteogenic growth factors secreted by hBMSCs into the local microenvironment can promote osteoblastic differentiation of hESCs, and the secretion of these factors was dependent on the state of cell differentiation.

MATERIALS AND METHODS

hESC Culture

The BG01 cell line was obtained from Bresagen, Inc. (Atlanta, GA) and cultured on irradiated mouse embryonic fibroblast (MEF) feeder layers at a density of approximately 19,000 cells/cm^2 onto 60 mm dishes (0.1% gelatin-coated). The hESC culture medium consisted of 80% (v/v) DMEM/F12, 20% (v/v) knockout serum replacement (KOSR), 200mM L-glutamine, 10mM nonessential amino acids (all obtained from Invitrogen), 14.3M β-mercaptoethanol (Sigma, St. Louis, MO), and 8 ng/ml bFGF (Invitrogen, Carlsbad, CA). Cell cultures were incubated at 37°C in 5% CO_2 in air and 95% humidity with medium changes everyday and manually passaged once per week. To induce osteogenic differentiation, the hESCs were made into embryoid bodies (EBs) and then seeded onto 0.1% gelatin-coated 6-well plates and cultured in hBMSC osteogenic medium (OS) consisting of 90% α-MEM (v/v), 10% heat-inactivated fetal bovine serum (FBS), 200 mM L-glutamine, and 10 mM nonessential amino acids (all obtained from Invitrogen) with 100 nM dexamethasone, 10 mM β-glycerophosphate, and 50 μM ascorbic acid for 4 weeks with medium changes every 48 hours [3].

hESC/hBMSC Transwell Co-Culture

The hBMSCs were isolated from patients at the University of Michigan using IRB approved protocols. The hBMSCs were plated at 2,000/cm^2 onto transwell inserts (0.4 μm pore, Corning, Corning, NY) and allowed to differentiate for 7-24 days in OS with complete medium changes every 48 hours. Two days prior to adding the hBMSC transwell inserts, the hESCs were either made into EBs or manually passaged directly from MEFs (omitting EB formation) and then seeded onto gelatin-coated 6-well plates. After 7, 14 or 24 days of differentiation, hBMSCs on transwell inserts were added to the 6-well plates and cultured in hBMSC growth medium (GM) without osteogenic supplements for an additional 28 days with medium changes every 48 hours. This method allows for the passage of soluble molecules while preventing direct cell-cell contact.

RT-PCR

Total RNA was obtained using Trizol (Invitrogen) and purified using the Qiagen RNEasy kit with DNase I treatment (Qiagen, Valencia, CA). Reverse transcription of 1 μg of RNA was performed using the Superscript III kit (Invitrogen). Taq DNA Polymerase (Invitrogen) was used to amplify the cDNA. The PCR conditions were as follows: 2 minutes at 94°C; followed by cycles of 45" denaturation at 94°C, 45" annealing at 56°C, and 60" extension at 72°C. Primer sequences were as follows (forward, reverse): *oct4* (GAA GGT ATT CAG CCA AAC, CTT AAT CCA AAA ACC CTG G) [11]; *nanog* (GAC TGA GCT GGT TGC CTC AT, TTT CTT CAG GCC CAC AAA TC) [12]; *runx2* (CAT GGT GGA GAT CAT CGC, ACT CTT GCC TCG TCC ACT C) [11]; *osterix* (GCA GCT AGA AGG GCG TGG TG, GCA GGC AGG TGA ACT TCT TC) [13]; *collagen1-Col-1* (GGA CAC AAT GGA TTG CAA GG, TAA CCA CTG CTC CAC TCT GG) [14]; *osteocalcin-OCN* (ATG AGA GCC CTC ACA CTC CTC, GCC GTA GAA GCG CCG ATA GGC) [14]; *GAPDH* (TGA AGG TCG GAG TCA ACG GAT TTG GT, CAT GRG GGC CAT GAG GTC CAC CAC) [12]. PCR products were analyzed on a 1.5% agarose gel with ethidium bromide staining. Imaging was obtained using a Fluor-S system (Biorad).

Mineralization Assays by Alizarin Red S and von Kossa Staining

Cell cultures were fixed in 4% paraformaldehyde for 30 minutes, washed twice with PBS and then stained. Alizarin Red S (Sigma) staining was used to determine the presence of mineralized nodules. Fixed cells were incubated for 1 hour in 1% Alizarin Red S solution and then washed twice with water. Von Kossa staining was used to determine the presence of phosphate. Fixed cells were incubated for 1 hour in 5% AgNO$_3$ solution and exposed to bright light for at least 30 minutes. For mineralization quantification, Alizarin Red S precipitate was extracted using a 10% acetic acid/20% methanol solution for 45 minutes. Spectrophotometric measurements of the extracted stain were made at 450 nm.

Immunofluorescence

Cell cultures were fixed for 30 minutes at room temperature with 4% paraformaldehyde. After washing with PBS, cells were permeabilized for 10 minutes with 10% (v/v) Triton X-100 in PBS, then washed twice with a serum wash containing 1% (v/v) sodium azide, followed by a blocking step containing 0.5% (v/v) Triton X-100 and 1% (v/v) sodium azide for 30 minutes at room temperature. Dilution buffer containing 2 μg/ml polyclonal rabbit anti-human osteocalcin (Santa Cruz Biotechnology, Santa Cruz, CA) was added and incubated at 4°C overnight. Following incubation, the cells were rinsed twice with serum wash and incubated in the dark with 8 μg/ml FITC-labeled secondary antibodies for 30 minutes and counterstained with DAPI. The cells were then washed with PBS and fluorescence was observed using a Nikon Eclipse TE3000.

Statistical Analysis

Results were evaluated using the student *t* test. Statistical significance was set at the 95% confidence level with a p-value < 0.05.

RESULTS

hESCs Undergo Osteogenic Differentiation and Give Rise to Condensed Mineralized Nodules in the Presence of hBMSCs

Human bone marrow stromal cells are able to differentiate into osteoblasts both *in vitro* and *in vivo*. Therefore, we postulated that exposing hESCs to hBMSCs undergoing osteoblastic differentiation would stimulate differentiation of hESCs along the osteogenic lineage [4-7]. For the transwell co-culture experiments, we used EB-derived cells. Human ESCs were plated onto gelatin-coated 6-well plates two days prior to co-culture with hBMSCs. Prior to plating the hESCs, hBMSCs were seeded onto gelatin-coated 0.4 μm pore transwell inserts and differentiated in growth medium plus defined osteogenic supplements (OS) for 7 days, 14 days, or 24 days. As controls, hESCs were co-cultured with undifferentiated hBMSCs or cultured alone in growth medium without osteogenic supplements (GM). At each specific time point (7, 14 or 24 days), the differentiated hBMSCs were added to the hESCs that had been plated onto gelatin-coated 6-well plates. The two cell types were then co-cultured in GM for an additional 28 days. Since the two cell types were not in direct physical contact, the hESCs were exposed to soluble signaling factors derived only from differentiated hBMSCs. In addition, hESCs were grown in osteogenic medium as previously described [1, 3].

Mineralized nodule formation is the hallmark of *in vitro* osteogenic differentiation. Therefore, after 28 days the cells and extracellular matrix (ECM) produced by co-cultures were stained by von Kossa and Alizarin Red to detect the presence of phosphate and calcium, respectively [15, 16]. The hESCs in co-culture with 14 day differentiated hBMSCs formed bone nodules that stained positively for von Kossa and Alizarin Red (Figure 1C, D). In contrast, cells grown with undifferentiated hBMSCs only showed minimal levels of mineral deposition and weak staining (Figure 1A, B). As expected, the hESCs grown in the presence of osteogenic supplements for 28 days stained positive for a mineralized matrix (data not shown).

Calcium deposition can also be quantified by extracting the Alizarin Red stain and subsequent spectrophotometric readings of Alizarin Red uptake. Therefore, we also performed quantitative Alizarin Red assays to assess the extent of bone nodule mineralization. There was a marked increase in mineralization within the transwell co-cultures with differentiated hBMSCs as compared to the controls. The hESCs exposed to hBMSCs differentiated for 7 days showed a 1.7-fold increase in Alizarin Red concentration (μg/ml) above control, a 2.5-fold increase for 14 day differentiated hBMSCs, and a 1.8-fold increase for cells exposed to hBMSCs differentiated for 24 days (Figure 2). Taken together, the upward trend of Alizarin Red content and greater von Kossa staining indicates that higher levels of mineral deposition could be found in hESCs exposed to differentiating hBMSCs.

hESCs Respond to the hBMSC Transwell Co-Culture System and Express Osteogenic Specific Markers

Osteoblastic lineage commitment can be observed by the expression of bone specific transcription factors runx2 and osterix, and with collagen type I production. Runx2 is a homolog of the *Drosophila* Runt protein that acts as a transcriptional regulator of osteoblast differentiation, and osterix is a zinc finger containing transcription factor required for osteoblastic differentiation that acts downstream of runx2 [17, 18]. Collagen type I is the most abundant protein found in bone ECM [19].

Therefore, in order to further confirm osteogenic differentiation, we analyzed the expression of these bone markers after 28 days in co-culture (Figure 3). RT-PCR analysis demonstrated that undifferentiated hESCs exhibit strong expression of the hESC pluripotency markers Oct-4 and Nanog, whereas they were absent in the transwell co-cultures and OS conditions (Figure 3, Lane 1).

In the presence of OS, there was a significant induction of osteogenic marker expression in hESCs [3, 15]. The transwell co-cultures also led to the induction of osteogenic gene expression, although it was not as robust. There were modest levels of bone specific gene expression by cells in co-culture with undifferentiated hBMSCs (Figure 3, Lane 4). However, gene expression of osteo-specific markers was upregulated in hESCs in the presence of differentiated hBMSCs (Figure 3, Lanes 5-7).

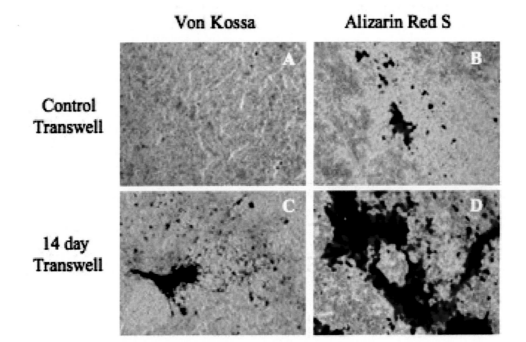

Figure 1. Human embryonic stem cells mineralize when co-cultured with differentiated hBMSCs. hBMSCs were grown on transwell inserts in osteogenic induction medium for 14 days. Differentiated hBMSCs were co-cultured with hESCs for an additional 28 days. Von Kossa (A, C) and Alizarin Red S (B, D) staining of hESCs co-cultured with differentiated or undifferentiated hBMSCs. Magnification = 10x.

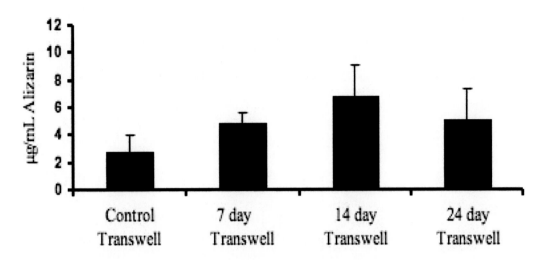

Figure 2. Quantitative analysis of Alizarin Red assays. hBMSCs were grown on transwell inserts in osteogenic induction medium for 7, 14 or 24 days. Differentiated hBMSCs were co-cultured with hESCs for an additional 28 days. Comparison of Alizarin Red staining between co-cultures with undifferentiated hBMSCs and experimental groups consisting of 7, 14 and 24 day differentiated hBMSCs.

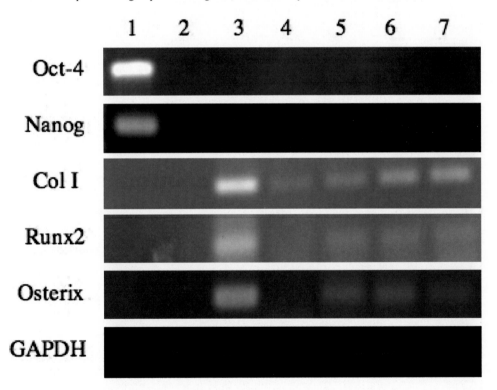

Figure 3. Human embryonic stem cells express osteoblastic markers when co-cultured with differentiated hBMSCs. hBMSCs were grown on transwell inserts in osteogenic induction medium for 7, 14 or 24 days. Differentiated hBMSCs were co-cultured with hESCs for an additional 28 days. Lane 1: undifferentiated hESCs; lane 2: hESC in growth medium; lane 3: hESC in osteogenic medium; lane 4: transwell co-culture with undifferentiated hBMSCs (control); lane 5: transwell co-culture with 7 day differentiated hBMSCs; lane 6: transwell co-culture with 14 day differentiated hBMSCs; lane 7: transwell co-culture with 24 day differentiated hBMSCs.

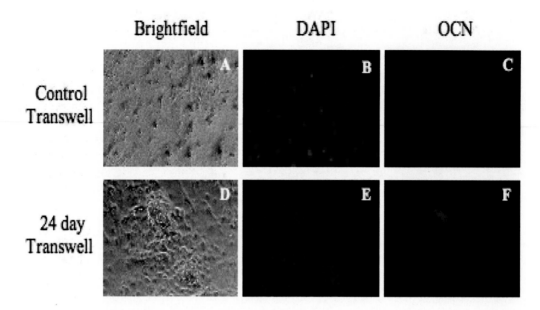

Figure 4. Human embryonic stem cells express osteocalcin when co-cultured with differentiated hBMSCs. hBMSCs were grown on transwell inserts in osteogenic induction medium for 24 days. Differentiated hBMSCs were co-cultured with hESCs for an additional 28 days. Expression of osteocalcin with DAPI counterstaining of hESC co-cultures with undifferentiated hBMSCs (A-C) and 24 day differentiated hBMSCs (D-F). Magnification = 10x.

hESCs Co-Cultured with Differentiated hBMSCs Secrete the ECM Protein Osteocalcin

Fully differentiated osteoblasts have the ability to form three-dimensional nodules with many of the immunohistochemical markers of bone and also form mineralized matrix. These nodules are thought to represent the end stage of differentation of osteoprogenitor cells *in vitro* [20]. It has also been shown that three-dimensional bone nodules derived from hESCs stain positive for osteocalcin (OCN) [15]. Results from our study show that immunostaining with antibody to OCN had strong immunoreactivity localized to clusters of hESCs co-cultured with 7, 14, and 24 day differentiated hBMSCs, whereas there was no evidence of OCN staining for hESCs exposed to undifferentiated hBMSCs (Figure 4). We believe the clusters that formed in the differentiated hBMSC transwell co-cultures are condensed bone nodules comprised of hESC-derived osteoblasts that secrete osteocalcin. The hESCs readily adhered to the culture plates and proliferated extensively in the presence of hBMSCs as evidenced by DAPI staining (Figure 4). As compared to hESCs grown alone or with undifferentiated hBMSCs, the cells in transwell co-culture with differentiated hBMSCs displayed bone nodule formation and mineralized matrix deposition.

DISCUSSION

Bone marrow is a complex tissue comprised of hematopoietic precursors, as well as a connective tissue network referred to as bone marrow stroma. It has been demonstrated that culture-adherent cells present in the marrow stroma have the capability to differentiate along multiple mesenchymal lineages [21]. Therefore, within the stroma itself exists a heterogeneous population of cells including osteoprogenitor cells that can proliferate and differentiate into mature osteoblasts. These multipotent cells are referred to as mesenchymal stem cells (MSCs), and have been serially passaged without lineage progression and subsequently shown to form cartilage, bone, fat and mature stromal cell lineages [2, 5, 22-24]. Due to our knowledge about the differentiation potential of human MSCs, we hypothesized that exposing hESCs to hBMSCs at various time points along the osteogenic differentiation pathway would contribute to directed osteoblastic differentiation of hESCs. The temporal expression of cell growth and osteoblast phenotype-related genes during osteoblast growth and differentiation has been established for primary osteoblasts, where three principle periods of osteoblast phenotype development have been identified – proliferation, ECM development and maturation, and mineralization phases [25]. During the proliferation phase (approximately 0-12 days in culture) collagen type I is predominantly expressed. The ECM development and maturation phase (approximately 12-21 days in culture) is characterized by high alkaline phosphatase expression, and within the mineralization phase (approximately 21-35 days in culture) osteocalcin is highly expressed as cell aggregates, or nodules, become mineralized in the presence of dexamethasone, ascorbic acid and β-glycerophosphate [25]. Therefore, we chose to differentiate our hBMSCs for 7, 14, and 24 days to determine if soluble factors secreted at different developmental time points would contribute to hESC differentiation.

Mineralization assays, RT-PCR and immunohistochemistry data described in this study show that by exposing the hESCs to soluble factors secreted by differentiated hBMSCs during each of the osteoblast phenotype development periods, we were able to achieve osteogenic differentiation without the addition of any other exogenous factors. The osteogenic response observed can be compared to that of hBMSCs in osteogenic medium [22]. There were substantially higher levels of Alizarin Red staining found in hESC co-cultures with differentiated hBMSCs than co-cultures with undifferentiated hBMSCs. More specifically, the cells exposed to hBMSCs differentiated for 14 days gave rise to highly mineralized bone nodules as evidenced by exhibiting the strongest von Kossa and Alizarin Red staining, as well as the highest Alizarin Red concentration as compared to the other transwell co-cultures. Although the increase in Alizarin Red content was not statistically significant between the control transwell and differentiated hBMSC transwell co-cultures, the 1.5-2.5 fold increase in concentration suggests that increased mineralization occurs when hESCs are exposed to soluble factors derived from differentiated hBMSCs. Additionally, the expression of bone-specific transcription factors, runx2 and osterix, as well as collagen type I, further confirmed that we successfully obtained hESC-derived osteoblasts. Gene expression of runx2, osterix, and collagen type I was found to be upregulated by co-culture conditions with differentiated hBMSCs, which is consistent with a mature osteoblast phenotype. There were low levels of bone specific gene expression within the control transwell co-culture with undifferentiated hBMSCS, and this is likely due to the fact that spontaneous differentiation occurs, thus yielding a small population of cells undergoing osteogenic differentiation. When the cells were cultured in osteogenic supplements for 28 days there was significantly stronger induction of bone specific gene expression as compared to the transwell co-cultures. This suggests that the growth factors secreted by the hBMSCs are osteo-inductive, however, further investigation regarding the identification and temporal expression of these secreted factors is needed to optimize this co-culture system. Lastly, immunostaining of condensed bone nodules within the differentiated hBMSC transwell co-cultures verified the presence of osteocalcin, a late marker of osteogenesis that corresponds with induction of mineralization [2]. Such differentiation was not observed in hESCs co-cultured with undifferentiated hBMSCs (Figure 4C).

Human embryonic stem cells hold promise for future regenerative medicine strategies. They were first derived from the inner cell mass of blastocyst stage embryos and have since been studied to determine their capacity to differentiate into all cell types [26, 27]. Osteoblastic differentiation of mouse, human, and monkey embryonic stem cells *in vitro* has previously been shown using the traditional osteogenic supplements dexamethasone, ascorbic acid and β-glycerophosphate [2, 28, 29]. Additionally, co-culture systems have been used to direct differentiation of both murine and human ES cells. Murine ES cells were shown to differentiate into osteoblasts using fetal murine osteoblasts in transwell co-culture, and hESCs were shown to differentiate into osteoblasts using direct co-culture with primary bone derived cells [28, 30]. Also, hESCs have been proven to have the capacity to differentiate into cartilage both *in vitro* and *in vivo* when exposed to primary chondrocytes in a transwell co-culture system [31].

This study differs from previous reports on the osteogenic differentiation of hESCs in two important ways: 1) Using an indirect transwell co-culture system; and 2) Controlling the state of differentiation of hBMSCs to induce hESC differentiation. The osteogenic response of the hESCs to the co-culture system may be explained by the fact that differentiated hBMSCs provide necessary osteo-inductive signals. Collectively, these data strongly support the hypothesis that the lineage progression of hESCs to osteoblasts can be directed by soluble signaling factors secreted by differentiated hBMSCs within the local cell environment. This co-culture system indicates the importance of cellular communication and coordination; therefore, determining the role of pro-osteogenic growth factors derived from hBMSCs and the influence of the microenvironment on osteogenic hESC differentiation is essential. Understanding the regulatory mechanisms that control osteoblastic differentiation will provide segue towards developing a useful cell source for bone tissue engineering repair. Future studies include performing proteomic analyses to identify and define the secreted factors, as well as systematically altering the microenvironment to further investigate the co-culture system we have developed and how it contributes to the osteogenic lineage progression of hESCs.

ACKNOWLEDGEMENTS

The authors would like to thank the University of Michigan Human Embryonic Stem Cell Core. This work was funded by the NIH National Institute for Dental and Craniofacial Research (DE016530) and (DE016530-01S1).

REFERENCES

[1]   Cao T, Heng B, Ye C, et al. Osteogenic differentiation within intact human embryoid bodies result in a marked increase in osteocalcin secretion after 12 days of in vitro culture, and formation of morphologically distinct nodule-like structures. *Tissue and Cell.* 2005;37:325-334.

[2]   Karp J, Ferreira L, Khademhosseini A, Kwon A, Yeh J, Langer R. Cultivation of human embryonic stem cells without the embryoid body step enhances osteogenesis in vitro. *Stem Cells.* 2006;24:835-843.

[3]   Sotille V, Thomson A, McWhir J. In vitro osteogenic differentiation of human ES cells. *Cloning and Stem Cells.* 2003;5:149-155.

[4]   Bianco P, Riminucci M, Gronthos S, Robey P. Bone Marrow Stromal Stem Cells: Nature, Biology, and Potential Applications. *Stem Cells.* 2003;19(3):180-192.

[5]   Jaiswal N, Haynesworth S, Caplan A, Bruder S. Osteogenic differentiation of purified, culture-expanded human mesenchymal stem cells in vitro. *Jour of Cell Biochem.* 1997;64:295-312.

[6]   Kuo C, RS T. Tissue engineering with mesenchymal stem cells. Tissue Eng in Med and Bio. 2003;22:51-56.

[7]   Wagers A, Weissman I. Plasticity of adult stem cells. *Cells.* 2004;116:639-648.

[8]   Beresford J, Graves S, Smoothy C. Formation of mineralized nodules by bone derived cells in vitro: a model of bone formation? *Amer. J. Med. Gen.* 1993;45:163-178.

[9]   Kalajzic I, Staal A, Yang W, et al. Expression profile of osteoblastic lineage at defined stages of differentiation. *J. Biol. Chem.* 2005;280:24618-24626.

[10] Krebsbach P, Kuznetov K, Emmons R, Rowe D, Robey P. Bone formation in vivo: comparison of osteogenesis by transplanted mouse and human marrow stromal fibroblasts. *Transplantation.* 1997;63:1059-1069.

[11] Bielby B, Boccaccini A, Polak J, Buttery L. In vitro differentiation and in vivo mineralization of osteogenic cells derived from human embryonic stem cells. *Tissue Eng.* 2004;10:1518-1525.

[12] Brimble S, Zen S, Weiler D, et al. Karyotypic stability, genotyping, differentiation, feeder-free maintenance, and gene expression sampling in three human embryonic stem cell lines derived prior to August 9, 2001. *Stem Cells and Dev.* 2004;13:585-596.

[13] Miura M, Grothos S, Zhao M, et al. SHED: Stem cells from human exfoliated deciduous teeth. *PNAS.* 2003;10:5807-5812.

[14] Noth U, Osyczka A, Tuli R. Multilineage mesenchymal differentiation potential of human trabecular bone-derived cells. *J. Orthop. Res.* 2002;20:1060-1069.

[15] Bonewald L, Harris S, Rosser J, et al. Von Kossa staining alone is not sufficient to confirm that mineralization in vitro represents bone formation. *Calcif Tissue Int.* 2003;72:537-547.

[16] Wergedal J, Baylink D. Distribution of acid and alkaline phosphate activity in undemineralized sections of the rat tibial diaphysis. Jour of Histochemistry and Cytochemistry. 1969;17:799-806.

[17] Ducy P, Zhang R, Geoffroy V, Ridall A, Karsenty G. Osf2/Cbfa1: A transcriptional activator of osteoblast differentiation. *Cell.* 1997;89:747.

[18] Nakashima K, de Crombrugghe B. Transcriptional mechanisms in osteoblast differentiation and bone formation. *Trends Genet.* 2003;19:458.

[19] Olsen B. Collagen it takes and bone it makes. Curr Biol. 1996;6:645-647.

[20] Purpura K, Aubin J, Zanstra P. Sustained in vitro of bone progenitors is cell density dependent. *Stem Cells.* 2004;22:39-50.

[21] Friendenstein A. Precursor cells of mechanocytes. *Int. Rev. Cytol.* 1976;47:327-355.

[22] Haynesworth S, Baber M, Caplan A. Characterization of the unique mesenchymal stem cell phenotype in vitro. *Trans Ortho. Res. Soc.* 1995;20:7-12.

[23] Johnstone B, Yoo J, Barry F. In vitro chondrogenesis of bone marrow-derived mesenchymal cells. *Trans Orthop. Res. Soc.* 1996;21:65.

[24] Pittenger M, Mackay A, Beck S. Human mesenchymal stem cells can be directed into chondrocytes, adipocytes or osteocytes. *Mol. Biol. Cell.* 1996;7:305a.

[25] Stein G, Lian J. Molecular mechanisms mediating proliferation/differentiation interrelationships during progressive development of the osteoblast phenotype. *Endocrine Reviews.* 1993;4:424-441.

[26] Odorico J, Kaufman D, Thomson J. Multilineage differentiation from human embryonic stem cell lines. *Stem Cells.* 2001;19:193-204.

[27] Thomson J, Itskovitz-Eldor J, Shapiro S, et al. Embryonic stem cell lines derived from human blastocysts. *Science.* 1998;282:1145-1147.

[28] Buttery L, Bourne S, Xynos J, et al. Differentiation of osteoblasts and in vitro bone formation from murine embryonic stem cells. *Tissue Eng.* 2001;7:89-99.

[29] Yamashita A, Takada T, Narita J, Yamamoto G, Torii R. Osteoblastic differentiation of monkey embryonic stem cells. *Stem Cells and Cloning.* 2005;7:232-237.

[30] Ahn S, Kim S, Park K, et al. Primary bone-derived cells induce osteogenic differentiation without exogenous factors in human embryonic stem cells. *Biochem Biophys. Res. Commun.* 2006;340(403-408).

[31] Vats A, Bielby R, Tolley N, et al. Chondrogenic differentiation of human embryonic stem cells: the effect of the micro-environment. *Tissue Eng.* 2006;12:1687-1697.

In: Developments in Stem Cell Research   ISBN: 978-1-60456-341-2
Editor: Prasad S. Koka   © 2008 Nova Science Publishers, Inc.

Chapter 2

Hyaluronic Acid Promotes Osteochondrogenic Differentiation from Mesenchymal Stem Cells in a Three-Dimensional Culture System

Kiyoshi Yoneno, Eiji Tanaka*,
Shigeru Ohno, Kotaro Tanimoto,
Kobun Honda, Nobuaki Tanaka,
Yu-Yu Lin, Yuki Tanne,
Satoru Okuma and Kazuo Tanne

Department of Orthodontics and Craniofacial Developmental Biology, Hiroshima
University Graduate School of Biomedical Sciences
Hiroshima 734-8553, Japan

ABSTRACT

A high expression of hyaluronic acid (HA) in precartilage condensation at the early stage of endochondral ossification indicates its importance in bone and cartilage development. However, it remains unclear whether HA is associated with bone development by promoting the differentiation from mesenchymal stem cells (MSCs). In this study, we examined the effects of HA on the differentiation of MSCs into osteoblasts and chondrocytes in a three-dimensional collagen gel culture system.

Human MSCs were cultured in HA-collagen hybrid gel (0.15% collagen, 0.5 mg/ml HA) and collagen mono gel (0.15% collagen). The cells were treated by either osteogenic differentiation medium (ODM) or chondrogenic differentiation medium (CDM). The

---

* **Address correspondence and reprint requests to**: Eiji Tanaka, Department of Orthodontics and Craniofacial Developmental Biology, Hiroshima University Graduate School of Biomedical Sciences, 1-2-3 Kasumi, Minami-Ku, Hiroshima 734-8553, Japan. Tel: +81-82-257-5686. Fax: +81-82-257-5687. E-mail: etanaka@hiroshima-u.ac.jp

HA-collagen gel maintained in ODM exhibited greater alkaline phosphatase activity, calcium deposition, and gene expression of bone markers than the collagen mono gel. The HA-collagen gel maintained in CDM had greater glycosaminoglycan deposition and gene expression of cartilage markers than the collagen mono gel.

These findings indicate that the addition of HA to collagen gel would be highly recommended for promoting the differentiation of MSC to osteoblasts and chondrocytes in three-dimensional condition.

Keywords: human mesenchymal stem cell; three-dimensional collagen gel; osteogenesis; chondrogenesis; hyaluronic acid

INTRODUCTION

Recently, Autogenous transplantation has been successfully performed to repair various hard tissue defects. However, the tissue damage and surgical invasion are often considered to be crucial problems to the patient. It has, thus far, been highly anticipated the development of a cell transplantation technique with a certain artificial scaffold, resulting in a bone or cartilage tissue engineering, for a more secure and safe tissue defect treatment. An anagenetic technique using mesenchymal stem cells (MSCs) has a good affinity between the organism and material, because MSCs can proliferate quickly and differentiate into various tissues [1]. In combination with a three-dimensional culture system, it seems to be possible to develop a tissue block or scaffold consisting of MSCs, which can alter itself to a tissue component equivalent to peripheral tissue. However, an appropriate scaffold has not yet been developed for a three-dimensional culture system of MSCs.

Collagen is the predominant protein in hard tissue, and easily metabolized in the general organs and tissues. In addition, collagen solution has a unique characteristic; i.e. it undergoes gelatinization by adjusting the concentration, pH and temperature [2]. Furthermore, collagen has the ability to promote ossification and subsequent bone formation [3]. Therefore, collagen is considered to be an efficient transplant material for tissue regeneration. Recently, we have established a three-dimensional collagen gel culture system to generate bone and cartilage from MSCs [4]. However, the system requires a long period to generate hard tissues from MSCs. Therefore, the development of additional materials that enable a quick differentiation of MSCs into chondrocytes and osteoblasts is required.

The high expression of hyaluronic acid (HA) during endochondral bone formation is indispensable to bone and cartilage development. HA is a highly and widely distributed over various tissues as a component of extracellular matrices. In fetal tissues, HA is particularly rich [5, 6], but the content decreases during development [6]. Interestingly, HA has been reported to promote the migration and proliferation of MSCs [7, 8], and is associated with the acceleration of wound healing [9]. Furthermore, HA binds to specific cell-surface receptors, such as CD44 and the receptor for HA-mediated motility (RHAMM), and also binds to other matrix molecules, such as collagen and proteoglycans [6, 10-12]. Thus far, we hypothesized that HA exerts certain influences on the differentiation process from MSCs and becomes an appropriate scaffold for the three-dimensional culture system of MSCs.

This study was conducted to examine the differentiation of human MSCs embedded in the HA-collagen hybrid gels into chondrocytes and osteoblasts, and to elucidate the usefulness of HA as an additional scaffold for the three-dimensional collagen gel culture system of MSCs.

MATERIALS AND METHODS

Cells and Three-Dimensional Collagen-HA Hybrid Gel

Human bone marrow MSCs and the growth medium (MSCGM) were purchased from Cambrex Bio Science Walkersville Inc. (Walkersville, MD). MSCs were cultured in HA-collagen hybrid gel composed of 0.15% collagen (Nitta Gelatin Inc., Osaka, Japan) in combination with 0.5 mg/ml 1.9 million molecular weight HA (Suvenyl®, Chugai Pharmaceutical Co., Tokyo, Japan). All cultures were maintained in MSCGM at 37°C in a humidified 5% CO_2 incubator.

Osteogenic and Chondrogenic Differentiation

For osteogenic differentiation, human MSCs were seeded in collagen-HA hybrid gel at a density of 5 x 10^5 cell/ml and cultured in the osteogenic differentiation medium (ODM), which was consisted of MSCGM supplemented with 100 nM dexamethasone (Dex), 10 mM β-glycerophosphate and 0.05 mM L-ascorbic acid-2-phosphate (AsAP) [13].

For chondrogenic differentiation, human MSCs were seeded in collagen-HA hybrid gel at a density of 5 x 10^6 cell/ml and cultured in the chondrogenic differentiation medium (CDM), which was consisted of serum-free MSCGM containing 1 mM sodium pyruvate, 100 μg/ml AsAP, 1 x 10^{-7} M Dex, 1% ITS, 5.33 μg/ml linolate, 1.25 mg/ml bovine serum albumin, 40 μg/ml proline and 10 ng/ml recombinant human TGF-β3 (RandD Systems, Minneapolis, MN) [14, 15].

As the controls, human MSCs cultured in 0.15% collagen gel without HA and treated with ODM or CDM were used.

Alizarin Red and Alkaline Phosphatase (ALP) Stainings

The gels cultured in ODM for 10 days were fixed with 100% ethanol and stained with 1% alizarin red (pH6.3) as previously described [16]. Alkaline phosphatase (ALP) staining was also performed for the gel cultures on day 5 using fast 5-bromo-4-chloro-3-indolyl phosphate/nitro blue tetrazolium tablets (Sigma-Aldrich Co., St. Louis, MO). The appearance of the cells in the gel was observed with phase contrast microscopy.

Measurement of DNA Content

DNA content in the gel was determined by using a PicoGreen dsDNA Quantitation kit (Molecular Probes, OR, USA). The gels were digested with papain solution (1 μg/ml papain in 50 mM sodium phosphate, pH6.5, containing 2 mM N-acetyl-L-cysteine and 2 mM EDTA) for 3 hours at 65°C, followed by fluorescein conjugation. Total DNA contents were quantified by reading the fluorescein at an excitation of 480 nm and an emission of 520 nm in a spectrofluorometer (Fluorescence spectrophoto-meter 850, Hitachi, Tokyo, Japan) with a reference to the fluorescein values obtained from the standard samples.

Calcium Incorporation Assay

The gels on day 14 were digested with 0.2% (v/v) triton-X-100, 0.02% collagenase and 6N HCl. The total amount of calcium was measured using a Calcium-C-kit (Wako, Osaka, Japan) and a microplate reader at an optimal density (OD) of 570 nm [17]. The levels of calcium were standardized for each well using DNA content per well.

ALP Activity Analysis

The gels on day 14 were digested with 0.2% (v/v) triton-X-100 and 0.02% collagenase, and incubated with 5 mM *p*-nitrophenyl phosphate in 50 mM glycine, 1 mM $MgCl_2$, pH10.5 at 37°C for 2 hrs [18, 19]. The ALP activity was estimated by quantifying the absorbance of *p*-nitrophenol product formed at an OD of 405 nm on a microplate reader. The levels of ALP activity were standardized for each well using DNA content per well.

Toluidine Blue Staining

The gels on day 10 were fixed with 100% ethanol and stained with 1% toluidine blue. To avoid the staining of HA, the staining procedure was performed at pH 2.5. Acidic muco-polysaccharide without HA could be found with pH 2.5 toluidine blue staining. The appearance of the cells in these gel cultures was observed with phase contrast microscopy.

Measurement of Sulfated Glycosaminoglycan (GAG) Contents

The gels on day 20 were digested with papain solution (1 μg/ml in 50 mM sodium phosphate, pH 6.5, containing 2 mM N-acetyl-L-cysteine, and 2 mM EDTA) for 3 hrs at 65°C [20]. The sulfated GAG contents, excluding HA, were quantified by the use of a Blyscan[TM] Sulfated GAG Assay kit (Biocolor Ltd., Northern Ireland, UK) and a microplate reader at an OD of 655 nm. The levels of sulfated GAG were standardized for each well using DNA content per well.

Semi-Quantitative Real Time Reverse Transcription Polymerase Chain-Reaction (Real Time RT-PCR) Analysis

Total RNA was isolated from the gel cultures using a guanidine thiocyanate method [21]. A single strand cDNA was synthesized from 1 µg of total RNA using Oligo $(dT)_{20}$ primer (Toyobo, Osaka, Japan) and a Rever Tra Ace-α first strand cDNA synthesis kit (Toyobo). The mRNA levels were determined by semi-quantitative real time RT-PCR analysis, using a SYBR Green PCR master mix (Applied Biosystems, Foster City, CA) or a TaqMan Universal PCR master mix (Applied Biosystems) with an automated fluorometer (ABI Prism 7700 sequence detection system; Applied Biosystems). Tables 1 and 2 show the sequences of the PCR primers and probes for type II collagen, type X collagen, aggrecan, Sox9, bone sialoprotein (BSP), ALP, type I collagen, Runx2, osterix and glyceraldehyde-3-phosphate dehydro-genase (GAPDH). The results of semi-quantitative real time RT-PCR analysis were assessed with a cycle threshold (Ct) value, which identifies a cycle when the fluorescence of a given sample becomes significantly different from the base signal. Quantification of the signals was performed by normalizing the signals of target genes relative to the GAPDH signals. Normalized Ct values were expressed relative to the controls.

Statistical Analysis

Experiments were performed at least in triplicate and expressed as mean ± standard deviations. The student's *t* test was used for statistical analysis.

Table 1. Primer Sequences Used in Chondrogenic Differentiation

Gene (Accession)	Primer sequence
Type II collagen (NM_001844)	5'-GGCAATAGCAGGTTCACGTACA-3' 5'-CGATAACAGTCTTGCCCCACTT-3'
Type X collagen (X60382)	5'-TCATGTTTGGGTAGGCCTGTA-3' 5'-GATCCAGGTAGCCTTTGGTGT-3'
Aggrecan (NM_001135)	5'-GCTACACAGGTGAAGACTTTGTGG-3' 5'-TTCACCCTCAGTGATGTTTCGAGG-3'
Sox9 (NM_000346)	5'-CAACGCCGAGCTCAGCAA-3' 5'-TCATGCCGGAGGAGGAGT-3'
GAPDH	5'-AAGGTGAAGGTCGGAGTCAAC-3' 5'-GAGTTAAAAGCAGCCCTGGTG-3'

Table 2. Primer and probe Sequences Used in Osteogenic Differentiation

Gene (Accession)	Primer and probe sequence
BSP (NM_004967)	Hs00173720_m1 (Applied Biosystems)
ALP (NM_001632)	Hs00740632_gH (Applied Biosystems)
Type I collagen (NM_000088)	Hs00164004_m1 (Applied Biosystems)
Runx2 (NM_001024630)	5'-TGGACGAGGCAAGAGTTTCA-3' 5'-ATACTGGGATGAGGAATGCG-3'
Osterix (AF477981)	5'-ATGGCGTCCTCCCTGCTTGA-3' 5'-TGCCCAGAGTTGTTGAGTCCCG-3'
GAPDH	Forward: 5'-AAGGTGAAGGTCGGAGTCAAC-3' Reverse: 5'-GAGTTAAAAGCAGCCCTGGTG-3' Probe: 5'-TTTGGTCGTATTGGGCGCCTGG-3'

* Sequences of oligonucleotide primers for BSP, ALP and Type I collagen purchased from Applied Biosystems are unknown.
** Runx2 and Osterix were analyzed with SYBR Green PCR. BSP, ALP, Type I collagen and GAPDH were analyzed with TaqMan Universal PCR.

RESULTS

Osteogenic Differentiation in the HA-Collagen Hybrid Gel

When the MSCs were stimulated by ODM for 5 days, the intensity of ALP staining was greater in the HA-collagen hybrid gel than in the collagen mono gel (Figure 1A). The ALP activity was also significantly two times greater in the HA-collagen hybrid gel than in the collagen mono gel (Figure 1B).

Alizarin red staining for the MSC gel culture showed enhanced calcium deposition in the extracellular regions of the gel cultures maintained in ODM. The intensity of the Alizarin red staining on day 10 was greater in the HA-collagen hybrid gel than in the collagen mono gel (Figure 2A). Furthermore, the incorporation of calcium into the extracellular matrix on day 14 was also significantly greater in the HA-collagen hybrid gel than in the collagen mono gel (Figure 2B).

The gene expressions of all bone markers examined, BSP, ALP, type I collagen, Runx2 and osterix were significantly greater in the HA-collagen hybrid gel than in the collagen mono gel (Figure 3). The relative increase in gene expression levels for BSP, ALP, type I collagen, Runx2 and osterix were 1.5-fold, 1.8-fold, 1.2-fold, 1.3-fold and 1.7-fold, respectively, as compared to those in the collagen mono gel.

Chondrogenic Differentiation in the HA-Collagen Hybrid Gel

To assess the chondrogenesis of MSCs in the three-dimensional gel, the gels were stained with toluidine blue after being cultured in CDM for 10 days. The intensity of the staining was greater in the HA-collagen hybrid gels than in the collagen mono gels (Figure 4A). Furthermore, the total amount of sulfated GAG in the collagen gels on day 20 was also significantly greater in gels containing HA than those without it (Figure 4B).

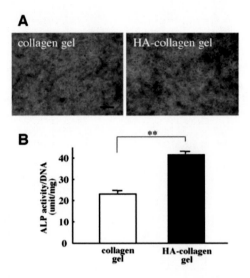

Figure 1. ALP activity in the HA-collagen gel culture of MSCs maintained in ODM. MSCs were seeded in the collagen and HA-collagen gels, and cultured in ODM for 14 days. The collagen and HA-collagen gels on day 5 were fixed and stained with ALP for 10 min (A). Bar indicates 50 μm. The gels on day 14 were digested and the ALP activity was determined (B). The data are the mean ± standard deviations of triplicate determinations. Statistical analyses were performed between the cultures with and without HA stimulation. **: p<0.01.

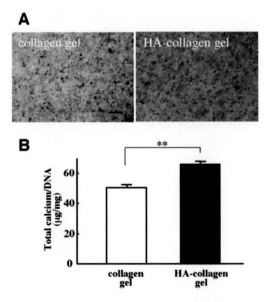

Figure 2. Mineralization in the HA-collagen gel culture of MSCs maintained in ODM. MSCs were seeded in the collagen and HA-collagen gels, and cultured in ODM for 14 days. The collagen and HA-collagen gels on day 10 were fixed and stained with alizarin red (A). Bar indicates 50 μm. The gels on day 14 were digested and the calcium contents were measured (B). The data are the mean ± standard deviations of triplicate determinations. Statistical analyses were performed between the cultures with and without HA stimulation. **: p<0.01.

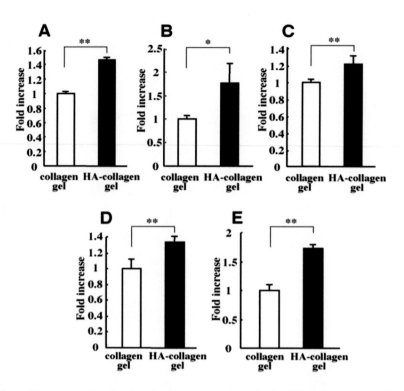

Figure 3. The mRNA expression levels of osteogenic markers in the HA-collagen gel culture of MSCs maintained in ODM. The mRNA expressions of BSP (A), ALP (B), type I collagen (C) and Runx2 (D) on day 5 and the mRNA expression of osterix (E) on day 14 were analyzed with semi-quantitative real time RT-PCR. Fold increases in the mRNA levels were determined in comparison to those in collagen mono gel. The data are the mean ± standard deviations of triplicate determinations. Statistical analyses were performed between the cultures with and without HA stimulation. *: p<0.05, **: p<0.01.

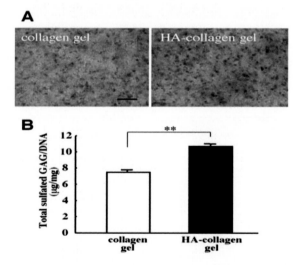

Figure 4. Sulfated GAG contents in the HA-collagen gel culture of MSCs maintained in CDM. MSCs were differentiated into chondrocytes in the collagen and HA-collagen gels for 20 days. The collagen and HA-collagen gels on day 10 were fixed and stained with pH2.5 toluidine blue for 1 min (A). Bar indicates 50 μm. The gels on day 20 were digested and the sulfated GAG contents were measured (B). The data are the mean ± standard deviations of triplicate determinations. Statistical analyses were performed for the cultures with and without HA stimulation. **: p<0.01.

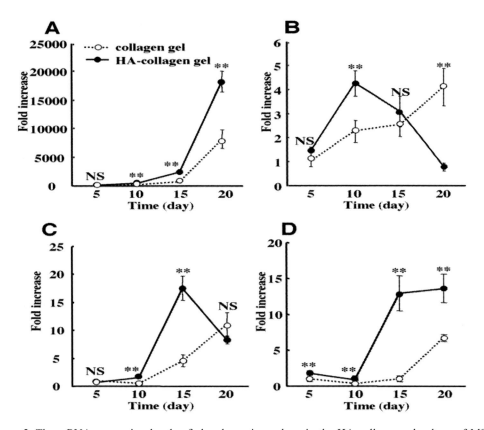

Figure 5. The mRNA expression levels of chondrogenic markers in the HA-collagen gel culture of MSCs maintained in CDM. MSCs were seeded in the collagen (opened circle) and HA-collagen (closed circle) gels cultured in CDM for 20 days. The mRNA expressions of type II collagen (A), type X collagen (B), aggrecan (C), and Sox9 (D) were analyzed with semi-quantitative real time RT-PCR. Fold increases in the mRNA levels of type II collagen, type X collagen, aggrecan, and Sox9 were determined in comparison to the levels on day 5 when maintained in the collagen gel without HA. The data are the mean ± standard deviations of triplicate determinations. Statistical analyses were performed for the cultures with and without HA stimulation on the same days. **: p<0.01 vs collagen gel, NS: not significant.

DISCUSSION

Various materials have been recently developed as scaffolds for three-dimensional culture. Collagen, which is one of the general materials for tissue engineering, has been used frequently as a matrix supportive component for the three-dimensional culture due to its easy gelatinization [2]. In addition, collagen has the ability to promote the differentiation of osteoblasts and subsequent bone formation [3]. These findings demonstrate that collagen can be easily and safely applied to clinical field. Although we have recently established the three-dimensional collagen gel culture system to develop the hard tissues from MSCs *in vitro* [4], it took two weeks or more to differentiate MSCs into osteoblasts and chondrocytes. The prompt differentiation from MSCs would be essential for clinical application, and it is absolutely necessary to find a method of quick differentiation from MSCs into osteoblasts and chondrocytes.

In this present study, we have first reported that the additional HA stimulates the promotion of MSCs' differentiation potential in comparison to non-HA gel. HA is ubiquitously distributed over the whole body and its structure is uniform within the various species, indicating a free-autoimmune response. Moreover, the metabolism of HA in the mammalian body is extremely quick compared to the other extracellular matrix components, due to several excellent digestion systems exerted by enzymes and receptors [22-25]. Thus, HA can be used conveniently and safely, even when using it for transplantation in a clinical aspect. In addition, our studies have shown that the osteogenic and chondrogenic differentiations from MSCs are markedly enhanced in the HA-collagen hybrid gel cultures compared to that in the collagen mono gels. These results suggest that HA is possibly a better candidate as a useful additional scaffold for the three-dimensional culture system for MSCs.

We adopted a high molecular weight HA (molecular weight of 1.9 million) at a concentration of 0.5 mg/ml in the gel in this investigation, because a high molecular weight HA at a concentration of 0.1 mg/ml to 1.0 mg/ml in medium has been reported to enhance the calcification in osteoblasts [26] and matrix production in chondrocytes [27, 28]. However, to evaluate the influence of the amount of collagen and HA on the tissue regeneration, more studies should be conducted in future.

It is assumed that HA is important in the differentiation process of the cell, because HA is rich in fetal tissues in particular, and because the expression of HA is upregulated in the process of cell differentiation [6]. Furthermore, HA can be found in almost all human cells, and is involved in wide variety of physiological and pathological processes, including wound healing, cell migration and cell proliferation [29]. The biomolecular mechanism of the effects of additional HA involved in MSCs' differentiation is poorly understood. Recently, CD44, the primary receptor for HA [29, 30], was reported to provide a physical link between cell and matrix [31, 32].

Although the signaling pathways by HA-CD44 interaction remains controversial, CD44 has, theoretically, the basic structure necessary to function in signal transduction. Therefore, the interaction of the HA and CD44 might strictly control proliferation and the differentiation of MSCs.

It was previously reported that a gel composed of collagen was contracted due to the influence of cell density and collagen concentration [33, 34], and we also found the similar extent of contractions between collagen mono gel and HA-collagen hybrid gel after long culture. For transplantation, a substantial volume of tissue is necessary; therefore, the next task will be focused on preventing such contraction of the gel after a long period of culture.

In conclusion, we have demonstrated that human MSCs had a higher differentiation potential to bone and cartilage in the HA-collagen hybrid gel than in the collagen mono gel. This implies that HA treatment has become a useful remedy for a more efficient and optimal hard tissue regeneration from MSCs.

ACKNOWLEDGMENTS

This study was supported by a Grant-in-aid (#15390636) for Scientific Research from the Ministry of Education, Science, Sports and Culture of Japan. This work was also carried out in courtesy of the Research Center for Molecular Medicine, Graduate School of Biomedical Sciences, Hiroshima University.

REFERENCES

[1]   Pittenger, MF; Mackay, AM; Beck, SC; Jaiswal, RK; Douglas, R; Mosca, JD; Moorman, MA; Simonetti, DW; Craig, S; Marshak, DR. Multilineage potential of adult human mesenchymal stem cells. *Science,* 1999, 284, 143-147.

[2]   Bell, E; Ivarsson, B; Merrill, C. Production of a tissue-like structure by contraction of collagen lattices by human fibroblasts of different proliferative potential in vitro. *Proc. Natl. Acad. Sci. USA,* 1979, 76, 1274-1278.

[3]   Celic, S; Katayama, Y; Chilco, PJ; Martin, TJ; Findlay, DM. Type I collagen influence on gene expression in UMR106-06 osteoblast-like cells is inhibited by genistein. *J. Endocrinol,* 1998, 158, 377-388.

[4]   Yoneno, K; Ohno, S; Tanimoto, K; Honda, K; Tanaka, N; Doi, T; Kawata, T; Tanaka, E; Kapila, S; Tanne, K. Multidifferentiation potential of mesenchymal stem cells in three-dimensional collagen gel cultures. *J. Biomed. Mater. Res. A,* 2005, 75, 733-741.

[5]   Agren, UM; Tammi, M; Ryynanen, M; Tammi, R. Developmentally programmed expression of hyaluronan in human skin and its appendages. *J. Invest. Dermatol.,* 1997, 109, 219-224.

[6]   Fraser, JR; Laurent, TC; Laurent, UB. Hyaluronan: its nature, distribution, functions and turnover. *J. Intern. Med,* 1997, 242, 27-33.

[7]   Pilloni, A; Bernard, GW. The effect of hyaluronan on mouse intramembranous osteogenesis in vitro. *Cell Tissue Res.,* 1998, 294, 323-333.

[8]   Toole, BP; Jackson, G; Gross, J. Hyaluronate in morphogenesis: inhibition of chondrogenesis in vitro. *Proc. Natl. Acad. Sci. USA,* 1972, 69, 1384-1386.

[9]   Oksala, O; Salo, T; Tammi, R; Hakkinen, L; Jalkanen, M; Inki, P; Larjava, H. Expression of proteoglycans and hyaluronan during wound healing. *J. Histochem. Cytochem,* 1995, 43, 125-135.

[10]   Underhill, C. CD44: the hyaluronan receptor. *J. Cell Sci,* 1992, 103 (Pt 2), 293-298.

[11]   Knudson, CB; Knudson, W. Hyaluronan-binding proteins in development, tissue homeostasis, and disease. *Faseb J.,* 1993, 7, 1233-1241.

[12]   Savani, RC; Wang, C; Yang, B; Zhang, S; Kinsella, MG; Wight, TN; Stern, R; Nance, DM; Turley, EA. Migration of bovine aortic smooth muscle cells after wounding injury. The role of hyaluronan and RHAMM. *J. Clin. Invest,* 1995, 95, 1158-1168.

[13]   Jaiswal, N; Haynesworth, SE; Caplan, AI; Bruder, SP. Osteogenic differentiation of purified, culture-expanded human mesenchymal stem cells in vitro. *J. Cell Biochem,* 1997, 64, 295-312.

[14]   Barry, F; Boynton, RE; Liu, B; Murphy, JM. Chondrogenic differentiation of mesenchymal stem cells from bone marrow: differentiation-dependent gene expression of matrix components. *Exp. Cell Res.,* 2001, 268, 189-200.

[15]   Sottile, V; Halleux, C; Bassilana, F; Keller, H; Seuwen, K. Stem cell characteristics of human trabecular bone-derived cells. *Bone,* 2002, 30, 699-704.

[16]   Stanford, CM; Jacobson, PA; Eanes, ED; Lembke, LA; Midura, RJ. Rapidly forming apatitic mineral in an osteoblastic cell line (UMR 106-01 BSP). *J. Biol. Chem.,* 1995, 270, 9420-9428.

[17]   Sugitani, H; Wachi, H; Murata, H; Sato, F; Mecham, RP; Seyama, Y. Characterization of an in vitro model of calcification in retinal pigmented epithelial cells. *J. Atheroscler. Thromb.,* 2003, 10, 48-56.

[18]   Piche, JE; Carnes, DL, Jr.; Graves, DT. Initial characterization of cells derived from human periodontia. *J. Dent Res.,* 1989, 68, 761-767.

[19]   Young, GP; Friedman, S; Yedlin, ST; Allers, DH. Effect of fat feeding on intestinal alkaline phosphatase activity in tissue and serum. *Am. J. Physiol.,* 1981, 241, G461-468.

[20]   Farndale, RW; Sayers, CA; Barrett, AJ. A direct spectrophotometric microassay for sulfated glycosaminoglycans in cartilage cultures. *Connect. Tissue Res.,* 1982, 9, 247-248.

[21]   Smale, G; Sasse, J. RNA isolation from cartilage using density gradient centrifugation in cesium trifluoroacetate: an RNA preparation technique effective in the presence of high proteoglycan content. *Anal. Biochem.*, 1992, 203, 352-356.

[22]   Kreil, G. Hyaluronidases--a group of neglected enzymes. *Protein Sci.*, 1995, 4, 1666-1669.

[23]   Frost, GI; Csoka, AB; Wong, T; Stern, R. Purification, cloning, and expression of human plasma hyaluronidase. *Biochem. Biophys. Res. Commun*, 1997, 236, 10-15.

[24]   Collis, L; Hall, C; Lange, L; Ziebell, M; Prestwich, R; Turley, EA. Rapid hyaluronan uptake is associated with enhanced motility: implications for an intracellular mode of action. *FEBS Lett*, 1998, 440, 444-449.

[25]   Tammi, R; Rilla, K; Pienimaki, JP; MacCallum, DK; Hogg, M; Luukkonen, M; Hascall, VC; Tammi, M. Hyaluronan enters keratinocytes by a novel endocytic route for catabolism. *J. Biol. Chem.*, 2001, 276, 35111-35122.

[26]   Huang, L; Cheng, YY; Koo, PL; Lee, KM; Qin, L; Cheng, JC; Kumta, SM. The effect of hyaluronan on osteoblast proliferation and differentiation in rat calvarial-derived cell cultures. *J. Biomed. Mater. Res.* A, 2003, 66, 880-884.

[27]   Kikuchi, T; Yamada, H; Shimmei, M. Effect of high molecular weight hyaluronan on cartilage degeneration in a rabbit model of osteoarthritis. *Osteoarthritis Cartilage*, 1996, 4, 99-110.

[28]   Kawasaki, K; Ochi, M; Uchio, Y; Adachi, N; Matsusaki, M. Hyaluronic acid enhances proliferation and chondroitin sulfate synthesis in cultured chondrocytes embedded in collagen gels. *J. Cell Physiol.*, 1999, 179, 142-148.

[29]   Aruffo, A; Stamenkovic, I; Melnick, M; Underhill, CB; Seed, B. CD44 is the principal cell surface receptor for hyaluronate. *Cell*, 1990, 61, 1303-1313.

[30]   Cichy, J; Pure, E. The liberation of CD44. *J. Cell Biol.*, 2003, 161, 839-843.

[31]   Knudson, CB; Knudson, W. Hyaluronan and CD44: modulators of chondrocyte metabolism. *Clin. Orthop. Relat Res*, 2004, S152-162.

[32]   Lisignoli, G; Grassi, F; Zini, N; Toneguzzi, S; Piacentini, A; Guidolin, D; Bevilacqua, C; Facchini, A. Anti-Fas-induced apoptosis in chondrocytes reduced by hyaluronan: evidence for CD44 and CD54 (intercellular adhesion molecule 1) invovement. *Arthritis Rheum*, 2001, 44, 1800-1807.

[33]   Grinnell, F. Fibroblast-collagen-matrix contraction: growth-factor signalling and mechanical loading. *Trends Cell Biol.*, 2000, 10, 362-365.

[34]   Galois, L; Hutasse, S; Cortial, D; Rousseau, CF; Grossin, L; Ronziere, MC; Herbage, D; Freyria, AM. Bovine chondrocyte behaviour in three-dimensional type I collagen gel in terms of gel contraction, proliferation and gene expression. *Biomaterials*, 2006, 27, 79-90.

In: Developments in Stem Cell Research
Editor: Prasad S. Koka

ISBN: 978-1-60456-341-2
© 2008 Nova Science Publishers, Inc.

Chapter 3

CELL CYCLE REGULATION IN MAINTAINING IMMORTALITY OF EMBRYONIC STEM CELLS

Rajarshi Pal, Ashish Mehta and Aparna Khanna [*]

Embryonic Stem Cell Group, Reliance Life Sciences Pvt. Ltd.
Dhirubhai Ambani Life Sciences Center
Rabale, Navi Mumbai-400 701, INDIA

ABSTRACT

Embryonic stem cells (ESC) unlike normal somatic cells and their adult counterparts can proliferate indefinitely in culture without compromising on their pluripotency. While maintaining their self renewing capacity during expansion, they do not appear to undergo the process of aging. Evidently, this phenomenon of escaping senescence in hESC is possible because of its high telomerase activity, cell cycle regulation, DNA repair mechanism, as well as the lack of cytogenetic, genomic, mitochondrial and epigenomic changes. However, this capacity in hESC is lost with the onset of differentiation into fully matured somatic cells. Thus, hESCs hold tremendous promise in cell replacement therapy for various degenerative disorders like Parkinson's and Alzheimer's disease, cardiac disorders, type-I diabetes and many more. Although the process of self renewal and senescence in hESC is known to be regulated by several signaling pathways, either independently or cohesively, the molecular mechanisms enabling this process are not clearly delineated. This article provides a summarized overview of the interplaying mechanisms known to govern self renewal in hESC, with special emphasis on the role of cell cycle regulation.

[*] **Author for Correspondence**: Telephone: +91-22-67678425. Fax: +91-22-67678099; E-mail: parna_khanna@relbio.com

INTRODUCTION

Chili Davis once said, "Growing old is mandatory; growing up is optimal"; nothing much has changed since then. Aging is still an inevitable fact for all multi-cellular organisms. The similar natural phenomenon ex vivo is termed as senescence. Although the mechanism of this phenomenon still remains poorly understood, testing of cells in culture has provided valuable insight into the cellular and molecular basis of aging. Primarily, there are two major hypotheses contributing to cell senescence (Zeng and Rao, 2006). The first hypothesis proposes that aging is a result of slow accumulation of damage that leads to cellular and finally tissue deterioration (Kirkwood and Austad, 2000). The second hypothesis suggests that aging is the result of a cellular programming that is controlled by a biological clock (de Magalhaes and Church, 2005; Prinzinger, 2005).

The cell cycle is a highly organized process that results in the duplication and transmission of genetic information from one generation to the next during cell division. It is well known that cancer cells tend to remain in the cell cycle oblivious to internal or external controls. Very similar is the case of most stem cell populations. To be more precise, stem cells are believed to have undergone premature loss over the control of replication with an obvious consequence of enhanced proliferation. So, cells grown in culture are expected to maintain a dynamic balance between replication and proliferation. Therefore, the margin of error being minimal, it is imperative to subject these cells for frequent testing in order to monitor for alterations in expression of potential regulators of senescence and aging.

Embryonic stem cells (ESCs) are unique to all stem cell populations, as they are able to escape the process of replicative senescence, and hence appear to be immortal in vitro. Evidences in both mouse ESCs and human ESCs have clearly shown their wondrous capacity of continued replication (Evans and Kaufman, 1981; Thomson et al., 1998). They maintain a normal karyotype, respond to growth factors and are able to give rise to mortal differentiated cells. But, ESCs unlike the carcinoma cells do not appear to be transformed.

This unique ability of ESCs projects them as a suitable model system to understand the process of aging. In this manuscript we would provide an overview of the regulatory aspects of cell cycle in the maintenance of self renewal and senescence in ESCs by reinterpreting reverential concepts and apprehension of new hypotheses.

ABSENCE OF SENESCENCE IN ES CELLS

The status of a cell, at which it appears to loose its proliferating capacity, is termed as "senescence". It is well accepted that cellular senescence is a result of changes in gene expression and or epigenetic modifications. For example, histone deacetylase inhibitors, which decondense chromatin and stimulate the transcription of some genes, can induce senescence- like state in human fibroblasts (Ogryzko et al., 1996). Other studies show that senescence is associated with subtle changes in the nuclear morphology and formation of a distinct chromatin structure, called senescence-associated hetero-chromatic foci (SAHF) (Narita et al., 2003). However, the most fundamental property of ESCs is that they can self renew indefinitely in culture (Thomson et al., 1998; Carpenter et al., 2004; Rosler et al., 2004). We have also been successfully growing our cell line, ReliCell®hES1 (Mandal et al.,

2006), over more than 60 passages which is equivalent to have undergone about 200 population doublings, without having developed any alterations in terms of genomic and epigenomic stability (Pal et al., 2007). Reports suggest that the capacity of hESCs to bypass senescence is not due to acquisition of genotypic abnormalities in long term propagation (Brimble et al., 2004). Furthermore, detailed single nucleotide polymorphism (SNP) analysis and mitochondrial DNA sequencing has demonstrated an overall remarkable stability of hESC (Maitra et al., 2005). It emerges therefore, that the lack of normal senescence in ESC is neither transient nor sporadic, but is truly inherent. Hence, it is understandable that there will be several factors which function in an organized fashion to maintain the self renewal capacity of the ESCs. Key differences in cell cycle control, regulation of telomerase expression and DNA repair has been identified by gene expression profiling of both mESCs and hESCs (Miura et al., 2004b; Ginis et al., 2004).

TRANSCRIPTIONAL FACTORS AND NETWORKS CONTROLLING SELF RENEWAL OF hES CELLS

Several key transcription factors have been implicated to play a pivotal role in maintenance of hESCs in an undifferentiated state. Among them, Oct3/4 and Nanog are perhaps the two most well-known homeobox transcription factors that are not only expressed both in mouse and human ESC, but are essential in maintaining pluripotency (Chambers and Smith, 2004; Zaehres et al., 2005). However, there is limited knowledge about the upstream regulators that mediate the transcription of these genes.

Oct3/4 (also known as POU5F1), which was initially identified in embryonic carcinoma (EC) cells (Rosner et al., 1990); is expressed particularly in ESCs, early embryos, and germ cells (Okamoto et al., 1990; Scholer et al., 1990). Further, Oct3/4- deficient embryos have been shown to die at peri-implantation stages of development (Nichols et al., 1998). The presence of Oct3/4 in human ESCs was reconfirmed, with differentiation of cells leading to its down-regulation (Reubinoff et al., 2000). Interestingly, STELLAR, a gene with similar expression to Oct3/4 has recently been identified (Avery et al., 2006). However, the functional role of this gene is yet to be established.

Nanog is also an important candidate of this elite group of transcription factors which are critical for the process of self-renewal in hESC. Nanog is (a NK2-family homeobox gene) also specifically expressed in pluripotent cells (Niwa et al., 2000; Chambers et al., 2003). Nanog null embryos at E5.5 show disorganization of extra-embryonic tissues with no discernible epiblast or extraembryonic ectoderm (Mitsui et al., 2003). Further, hESC overexpressing Nanog can be maintained in feeder-free system, even without the requirement of conditioned medium (Darr et al., 2006). More interestingly, transcription of Nanog is regulated by binding of Oct3/4 and Sox2 (Sry-related transcription factor) to the Nanog promoter, as confirmed by mutagenesis and in vitro binding assays (Rodda et al., 2005).

Several extrinsic signals such as LIF, BMP, FGF and Wnt can support self renewal of ESC through regulating the pluripotent genes as reviewed by Okita and Yamanaka, Pan and Thomson (Okita and Yamanaka, 2006; Pan and Thomson, 2007). Using microarray, we have shown that key markers associated with ES cell pluripotency are also highly expressed in ReliCell®hES1 (Table 1). It is likely that the activity of these main transcription factors like

Oct3/4, Nanog and Sox2 comprising this core transcriptional regulatory circuitry is added to by the presence of other co-factors and post-translational modifications (Boyer et al., 2005). However, the precise mechanisms by which these factors engage in cross talk with one another resulting in governance of ESC self-renewal remain elusive.

CELL CYCLE REGULATION IN BYPASSING SENESCENCE

To proliferate, cells transverse the cell cycle in several, discrete, well-controlled phases and any break down in this regulation may result into uncontrolled growth. The first is the G1 phase, where the cells commit to enter the cell cycle and prepare to duplicate their DNA in S phase. After S phase, cells enter the G2 phase, where DNA repair may take place along with preparation for mitosis in M phase. In the M phase, chromatids and daughter cells separate, after which, the cells can enter G1 or G0, a quiescent phase. Entry into each phase of the cell cycle is carefully regulated by receptor collectives, termed cell cycle checkpoints. At this stage, the cell is ready for responding to these external stimuli, communicated through a cascade of intracellular phosphorylations, by up-regulating expression of the cyclins and cyclin dependent kinases (CDKs) leading to cell cycle regulation. The Cyclin is the regulatory unit and CDK is its catalytic partner. Cyclins, with their bound and activated CDKs, function during distinct stages of the cell cycle.

Stem cells are defined by both their ability to make more replicas of themselves, a property known as "self renewal", and their ability to produce differentiated cells. It is more or less well established that asymmetric cell division is a defining characteristic of stem cells that enable them to simultaneously perpetuate themselves in culture (Morrison and Kimble, 2006). However, to understand self renewal, it is not sufficient merely to delineate how stem cell proliferation is controlled, because not all cell divisions involve self renewal. Are there specific signals that exert a combined effect on cell cycle regulation and maintenance of stem cell state? Or are proliferation and maintenance of stem cell state regulated independently by distinct signals? If yes, what are the appropriate downstream regulators starting from activation of a specific pathway? All these issues are critical as answers to these questions may provide important clues about how to induce ESCs into different lineages in a controlled manner in culture, an essential element in their therapeutic application.

Table 1. Expression of pluripotency related genes in hESCs

Category	Gene Symbol	Description	ReliCell®hES1
Pluripotency	Oct-3/4	POU domain, class 5, transcription factor 1	240
	Nanog	NK2-family homeobox gene	21
	DNMT 3ß	DNA methyl transferase 3ß	42
	UTF1	Undifferentiated embryonic stem cell transcriptional factor 1	149
	TDGF1	Teratocarcinoma-derived growth factor 1	1722
	Rex	Zinc figure protein 42	530

Table 1. (Continued)

Category	Gene Symbol	Description	ReliCell®hES1
	Lefty A	left-right determination, factor A	1136
	Lefty B	left-right determination, factor B	2635
	Dppa5	Developmental pluripotency associated 5	269
	TERF1	Telomeric repeat binding factor 1	1111
	TERF2	Telomeric repeat binding factor 2	735
	CX43	Gap junction molecule, Connexin 43	1589
	Lin28	Lin-28 homolog (C. elegans)	2540
	Podxl	Podocalyxin-like transcript variant	1063
	Gal	Galanin	1158
	BMP3B	Bone morphogenetic protein 3B	254
	Wnt3A	wingless-type MMTV integration site family, member 3A	296
	Wnt4	wingless-type MMTV integration site family, member 4	50
	LIFR	Leukemia inhibitory factor-receptor	ND
	gp130	Signal transducer and activator of transcription 3, glycoprotein 130	ND
	FoxD3	forkhead box D3	21

ND: Not Detected; values mentioned in column 4 are arbitrary intensity units as detected by Illumina bead array system (Pal et al., 2007).

p53

p53 is a tumor suppressor, recessive gene. p53 mutations- most commonly missense mutations disrupt p53-DNA binding, resulting into malignant transformation. Thus, function of p53 in cell cycle regulation is to help the cell in traversing through the cell cycle if there is DNA damage, either by halting the cell at a checkpoint till DNA repair mechanisms are activated, or by compelling the cell to undergo apoptosis if the damage is irreversible (Di Leonardo et al., 1994).

Rb

The retinoblastoma gene product (Rb) is another tumor suppressor gene that prevents the cell cycle from progressing when DNA damage has taken place (Classon and Harlow, 2002; Zheng and Lee, 2002). In parallel with p53, malignant transformation requires loss of function of both copies of the gene. The Rb protein essentially inhibits the cell from entering S phase of the cell cycle by binding to a transcription factor called E2F, which in turn prevents the downstream events such as promoter binding and transcription of c-myc and c-fos (Zeng and Rao, 2006).

It has been reported earlier that inactivation of p53 and Rb tumor suppressor genes facilitate cells to bypass senescence (Lin et al., 1998; Sherr, 1998), indicating that these pathways are instrumental in long term survival without undergoing senescence. Furthermore, inhibition of p53 and Rb modulates how a cell reciprocates to telomeric changes, DNA damage and other cell cycle regulators (Zeng and Rao, 2006; Figure 1). In the context of ESCs, it has been demonstrated by massively parallel signature sequencing (MPSS) and microarray analysis that there lies a direct correlation between prolonged self renewal capacity of hESCs and the differential expression of p53 and Rb pathway related genes (Brandenberger et al., 2004; Miura et al., 2004a; Pal et al., 2007) (Table 2). For example, RB1, E2F1, TP53, MDM2 are absent while CCNA1, CEB1, MYC, CROC4, DDIT4 are up-regulated in hESCs (Table 2). Likewise in mESCs, Rb is not expressed whereas several MCM proteins like MCM2, MCM3 and MCM7 are strongly expressed (Ginis et al., 2004). However, unlike hESCs, high level of p53 has been observed in mESCs (Ginis et al., 2004). Overall, these findings provide substantial evidence that p53 and Rb pathways are critical determinants of immortality in ESCs.

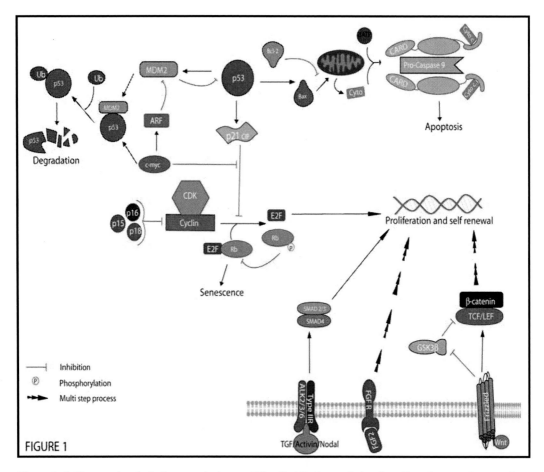

FIGURE 1

Figure 1. Self renewal and pluripotency in human ES cells. The key switches in cell cycle control are cyclins, CDKs and their downstream targets, p53 and Rb tumor suppressors. Regulation of p53 determines if a cell undergoes apoptosis or not. Apoptosis is cytochrome-C dependent, via activation of Bax and formation of caspase cascade. Conversely, several pathways are implicated in hESC self renewal. Wnt signaling is mediated by β-catenin/TCF/LEF via GSK-3β inhibition. TGF/activin/Nodal activates Smad-2/3, which associates with co-Smad-4 and translocates to the nucleus to activate ESC specific transcription factors. FGF signaling promotes the TGF/activin/Nodal pathway thereby inhibiting BMP pathway. However, this diagram does not preclude the role of other signaling molecules such as S1P, P13 kinase, PDGF and Neutrophins.

Table 2. Expression of self renewal related genes in hESCs

Category	Gene Symbol	Description	ReliCell®hES1
Self renewal and cell cycle regulation	NFIX	nuclear factor I/X (CCAAT-binding transcription factor)	107
	ZNF503	Zinc finger protein 503	580
	GPR56	G-protein couple receptor 56	509
	NR2F1	nuclear receptor subfamily 2, group F, member 1	359
	DDIT4	DNA-damage-inducible transcript 4	1217
	PEG3	paternally expressed 3	889
	SRGAP2	SLIT-ROBO Rho GTPase activating protein 2	516
	CDKN2B (p15)	Cyclin-dependent kinase inhibitor 2B	126.5
	TP53	Tumor suppressor protein, p53	ND
	MDM2	P53 binding protein	ND
	RB1	Retinoblastoma 1	ND
	E2F1	E2F transcriptional factor	ND
	CROC4	transcriptional activator of the c-fos promoter	389
	MYC	c-myc myelocytomatosis viral oncogene homolog (avian)	311
	CEB1	cyclin-E binding protein 1	792.5
	CCNA1	cyclin A1	76
Apoptosis	TNFRSF19	tumor necrosis factor receptor superfamily, member 8	418
	CASP3α	caspase 3, apoptosis-related cysteine protease, ß transcript	9
	CASP3β	caspase 3, apoptosis-related cysteine protease, ß transcript	592
	CARD11	caspase recruitment domain family, member 11	115
	ASC	apoptosis-associated speck-like protein containing a CARD	298
	PDCD8	programmed cell death 8 (apoptosis-inducing factor)	63

ND: Not Detected; values mentioned in column 4 are arbitrary intensity units as detected by Illumina bead array system (Pal et al., 2007).

OTHER CHANGES IMPLICATING IMMORTALITY OF HUMAN ES CELLS

Human ES cells in culture are prone to acquire DNA damage and likely undergo random mutations (Maitra et al., 2005). There is considerable data to suggest that hESCs are remarkably stable and certainly have in-built mechanisms to maintain their DNA stability in long term culture, which may work in concert with telomere regulation to allow cells to escape senescence (Brandenberger et al., 2004; Miura et al., 2004a; Maitra et al., 2005; Pal et al., 2007). Furthermore, many replication and repair-related genes have been shown to be highly expressed in hESCs implicating a high DNA repair activity (Miura et al., 2004a).

One of the commonest forms of epigenetic modifications in human cancers and aging tissues is DNA methylation at the CpG islands (where cytosine lies next to guanine), which may inhibit binding of transcription factors to suitable promoters (Razin and Kantor, 2005). Hence, monitoring DNA methylation patterns may help to determine if hESCs are changing in culture and are thus becoming vulnerable to senescence signals. Owing to its significance, the trend of examining methylation profiles of hESC lines in long term cultures has just begun (Maitra et al., 2005; Rugg-Gunn at al., 2005; Bibikova et al., 2006). However, these results warrant further analysis to establish a direct link between epigenetic status and immortality of hESCs.

Due to the scarcity of suitable repair mechanisms in the mitochondrial genome, it is more prone to DNA damage than its nuclear counterpart (Sherratt et al., 1997). Since, mitochondrial DNA mutations appear to play a role in the aging process (Wang et al., 2001), mitochondrial changes may be associated with cell cycle regulation leading to bypassing senescence. Although, it is premature to conclude, but there are a few reports on mitochondrial DNA analysis of hESCs (Maitra et al., 2005; Pal et al., 2007).

CONCLUSION

It is clearly evident that hESCs in culture sparks the activation of a number of signaling pathways, which seem to exert either independent or cooperative control to maintain its core stability and self renewal. Not surprisingly, cell cycle regulation appears well-defined, coupled with the presence of high levels of DNA repair enzymes and telomerase activity. Some of these mechanisms may facilitate hESCs to bypass senescence, although most of these pathways appear to be different from cancer cells and adult stem cells. Further, hESCs seem to maintain a unique genomic and epigenomic signature along with its mitochondrial DNA being stable over long term culture. These adjustments made by the hES cells suggest that these pathways contribute to sustain prolonged self renewal thus facilitating them to escape senescence and become immortal. However, it is noteworthy that the increasing trend of adopting feeder-free protocols with defined medium to grow hESC lines, may evoke different signaling pathways in maintaining ES cell identity. Factors that are essential in one culture system may be dispensable in another. Therefore, it becomes critical to compare and contrast the findings from various laboratories to identify the common elements of hESC culture and differentiation.

REFERENCES

Avery, S; Inniss, K; Moore, H. The regulation of self-renewal in human embryonic stem cells. *Stem Cell Dev.* 2006, 15, 729-740.

Bibikova, M; Chudin, E; Wu, B; Zhou, L; Garcia, EW; Liu, Y; Shin, S; Plaia, TW; Auerbach, JM; Arking, DE; Gonzalez, R; Crook, J; Davidson, B; Schulz, TC; Robins, A; Khanna, A; Sartipy, P; Hyllner. J; Vanguri, P; Savant-Bhonsale, S; Smith, AK; Chakravarti, A; Maitra, A; Rao, M; Barker, DL; Loring, JF; Fan, JB. Human embryonic stem cells have a unique epigenetic signature. *Genome Res.*, 2006, 16, 1075-1083.

Boyer, LA; Lee, TI; Cole, MF; Johnstone, SE; Levine, SS; Zucker, JP; Guenther, MG; Kumar, RM; Murray, HL; Jenner, RG; Gifford, DK; Melton, DA; Jaenish, R; Young, RA. Core transcriptional regulatory circuitry in human embryonic stem cells. *Cell*, 2005, 122, 947-956.

Brandenberger, R; Khrebtukova, I; Thies, RS; Miura, T; Jingli, C; Puri, R; Vasicek, T; Lebkowski, J; Rao, MS. MPSS profiling of human embryonic stem cells. BMC Dev Biol., 2004, 40, 10.

Brimble, SN; Zeng, X; Weiler, DA; Lou, Y, Liu, Y; Lyons, IG; Freed, WJ; Robin, AJ; Rao, MS; Schulz, TC. Karyotypic stability, genotyping, differentiation, feeder-free maintenance, and gene expression sampling in three human embryonic stem cells derived prior to August 9, 2001. *Stem Cell Dev.*, 2004, 13, 585-597.

Carpenter, MK; Rosler, ES; Fisk, GJ; Brandenberger, R; Ares, X; Miura, T; Lucero, M; Rao, MS. Properties of four human embryonic stem cell lines maintained in feeder-free culture system. *Dev. Dyn.*, 2004, 229, 243-258.

Chambers, I; Colby, D; Robertson, M; Nichols, J; Lee, S; Tweedie, S; Smith, A. Functional expression cloning of nanog, a pluripotency sustaining factor in embryonic stem cells. *Cell*, 2003, 113, 643-655.

Chambers, I; Smith, A. Self-renewal of teratocarcinoma and embryonic stem cells. *Oncogene,* 2004, 23, 50-60.

Classon, M; Harlow, E. The retinoblastoma tumor suppressor in development and cancer. *Nat. Rev. Cancer,* 2002, 2, 910-917.

Darr, H; Mayshar, Y; Benvenisty, N. Overexpression of NANOG in human ES cells enables feeder-free growth while inducing primitive ectoderm features. *Development,* 2006, 133, 1193-1201.

de Magalhaes, JP; Church, GM. Genomes optimize reproduction: aging as a consequence of the developmental program. *Physiology* (Bethesda), 2005, 20, 252-259.

Di Leonardo, A; Link, SP; Clarkin, K; Wahl, GM. DNA damage triggers a prolonged p53-dependent G1 arrest and long term induction of Cip1 in normal human fibroblast. *Genes Dev.*, 1994, 8, 2540-2551.

Evans, MJ; Kaufman, MH. Establishment in culture of pluripotential cells from mouse embryos. *Nature,* 1981, 292, 154-156.

Ginis, I; Luo, Y; Miura, T; Thies, S; Brandenberger, R; Gercht-Nir, S; Amit, M; Hoke, A; Carpenter, MK; Itskovitz-Eldor, J; Rao, MS. Differences between human and mouse embryonic stem cells. *Dev. Biol.*, 2004, 269, 360-380.

Kirkwood, TB; Austad, SN. Why do we age? *Nature,* 2000, 408, 233-238.

Lin, AW; Barradas, M; Stone, JC; van Aelst, L; Serrano, M; Lowe, SW. Premature senescence involving p53 and p16 is activated in response to constitutive MEK/MAPK mitogenic signaling. 1998. *Genes Dev.*, 12, 3008-3019.

Maitra, A; Arking, DE; Shivapurkar, N; Ikeda, M; Stastny, V; Kassauei, K; Sui, G; Cutler, DJ; Liu, Y; Brimble, SN; Noaksson, K; Hyllner, J; Schulz, TC; Zeng, X; Carpenter, M; Gazdar, AF; Rao, MS; Chakravati, A. Genomic alterations in cultured human embryonic stem cells. *Nat. Genet.*, 2005, 37, 1099-1103.

Mandal, A; Tipnis, S; Pal, R; Ravindran, G; Bose, B; Patki, A; Rao, MS; Khanna, A. Characterization and in vitro differentiation potential of a new human embryonic stem cell line, ReliCell®hES1. *Differentiation*, 2006, 74, 81–90.

Mitsui, K; Tokuzawa, Y; Itoh, H; Segawa, M; Murakami, M; Takahashi, K; Maruyama, M; Maeda, M; Yamanaka, S. The homeoprotein nanog is required for maintenance of pluripotency in mouse epiblast and ES cells. *Cell,* 2003, 113, 631-642.

Miura, T; Luo, Y; Khrebtukova, I; Brandenberger, R; Zhou, D; Thies, RS; Vasicek, T; Young, H; Lebkowski, J; Carpenter, MK; Rao, MS. Monitering early differentiation events in human embryonic stem cells by massive parallel signature sequencing and expressed sequence tag scan. *Stem Cell Dev.* 2004a, 13, 694-715.

Miura, T; Mattson, MP; Rao, MS. Cellular life span and senescence signaling in embryonic stem cells. *Aging Cell,* 2004b, 3, 333-343.

Morrison, SJ; Kimble, J. Asymmetric and symmetric stem cell division in development and cancer. *Nature,* 2006, 441, 1068-1074.

Narita, M; Nunez, S; Heard, E; Narita, M; Lin, AW; Hearn, SA; Spector, DL; Hannon, GJ; Lowe, SW. Rb-mediated heterochromatin formation and silencing of E2F target genes during cellular senescence. *Cell,* 2003, 113, 703-716.

Nichols, J; Zevnik, B; Anastassiadis, K; Niwa, H; Klewene-benius, D, Chambers, I; Scholer, H; Smith, A. Formation of pluripotent stem cells in the mammalian embryo depends on the POU transcription factor Oct 4. *Cell,* 1998, 95, 379-391.

Niwa, H; Miyazaki, J; Smith, AG. Quantitative expression of Oct-3/4 defines differentiation, dedifferentiation or self-renewal of ES cells. *Nat. Genet.,* 2000, 24, 372-376.

Ogryzko, V; Hirai, T; Russanova, V; Barbie, D; Howard, B. Human fibroblast commitment to a senescence-like state in response to histone deacetylase inhibitors in cell cycle dependent. *Mol. Cell Biol.,* 2006, 16, 5210-5218.

Okamoto, K; Okazawa, H; Okuda, A; Sakai, M; Muramatsu, M; Hamada, H. A novel octamer binding transcriptional factor is differentially expressed in mouse embryonic cells. *Cell,* 1990, 60, 461-473.

Okita, K; Yamanaka, S. Intracellular signaling pathways regulating pluripotency of embryonic stem cells. Curr *Stem Cell Res. Ther.,* 2006, 1, 103-111.

Pan, G; Thomson, JA. Nanog and transcriptional networks in embryonic stem cell pluripotency. *Cell Res.,* 2007, 17, 42-49.

Pal, R; Mandal, A; Rao, SH; Rao, MS; Khanna, A. A panel of tests to standardize characterization of human embryonic stem cells. *Reg. Med.,* 2007, in press.

Prinzinger, R. Programmed ageing: the theory of maximal metabolic scope. How does the biological clock tick? *EMBO Rep.,* 2005, 6, S14-S19.

Razin, A; Kantor, B. DNA methylation in epigenetic control of gene expression. *Prog. Mol. Subcell Biol.,* 2005, 38, 151-167.

Reubinoff, BE; Pera, MF; Fong, CY; Trounson, A; Bongso, A. Embryonic stem cell lines from human blastocysts: somatic differentiation in vitro. *Nat. Biotechnol.,* 2000, 18, 399-404.

Rodda, DJ; Chew, JL; Lim, LH; Loh, YH; Wang, B; Ng, HH; Robson, P. Transcriptional regulation of nanog by OCT4 and SOX2. *J. Biol. Chem.,* 2005, 280, 24731-24737.

Rosler, ES; Fisk, GJ; Ares, X; Irving, J; Miura, T; Rao MS; Carpenter, MK. Long-term culture of human embryonic stem cells in feeder-free conditions. *Dev. Dyn.,* 2004, 229, 259-274.

Rosner, MH; Vigano, MA; Ozato, K; Timmons, PM; Poirier, F; Rigby, PW; Staudt, LM. A POU-domain-transcription factor in early stem cell and germ cells of the mammalian embryo. *Nature,* 1990, 345, 686-692.

Rugg-Gunn, PJ; Ferguson-Smith, AC; Pedersen, RA. Epigenetic status of human embryonic stem cells. *Nat. Genet.,* 2005

Scholer, HR; Ruppert, S; Suzuki, N; Chowdhury, K; Gruss, P. A new type of POU domain in germ line-specific protein Oct-4. *Nature,* 1990, 344, 435-439.

Sherr, CJ. Tumor surveillance via the ARF-p53 pathway. 1998. *Genes Dev.,* 12, 2984-2991.

Sherratt, EJ; Thomas, AW; Alcolado, JC. Mitochondrial DNA defects: A widening clinical spectrum of disorders. *Clin. Sci.,* 1997, 92, 225-235.

Thomson, JA; Itskovitz-Eldor, J; Shapiro, SS; Waknitz, MA; Wiergiel, JJ; Marshall, VS; Jones, JM. Embryonic stem cell lines derived from human blastocysts. *Science,* 1998, 282, 1145-1147.

Wang, Y; Mickhkawa, Y; Mallidis, C; Bai, Y; Woodhouse, L; Yarasheski, KE; Miller, CA; Kskanas, V; Engel, WK; Bhasin, S; Attardi, G. Mouse specific mutations accumulate with aging in critical human mtDNA control sites for replication. *Proc. Nat. Acad. Sci. USA,* 2001, 98, 4022-4027.

Zaehres, H; Lensch, MW; Daheron, L; Stewart, SA; Itskovitz-Eldor, J; Daley, GQ. High-efficiency RNA interference in human embryonic stem cells. *Stem Cells,* 2005, 23, 299-305.

Zeng, X; Rao, MS. Human embryonic stem cells: long term stability, absence of senescence and a potential cell source for neural replacement. *Neuroscience,* 2006, doi:10.1016/j.neuroscience.2006.09.017.

Zheng, L; Lee, WH. Retinoblastoma tumor suppressor and genome stability. *Adv. Cancer Res.,* 2002, 85, 13-50.

In: Developments in Stem Cell Research
Editor: Prasad S. Koka

ISBN: 978-1-60456-341-2
© 2008 Nova Science Publishers, Inc.

Chapter 4

POTENTIAL TARGET MOLECULES FOR *EX VIVO* EXPANSION OF HEMATOPOIETIC STEM CELLS AND THEIR ROLES IN NORMAL HEMATOPOIESIS

Hirokazu Tanaka[*1, 2], *Itaru Matsumura*[*2] *and Yuzuru Kanakura*[*2]

[1] Department of Regenerative Medicine, Institute of Biomedical Research and Innovation, Kobe, Japan
[2] Department of Hematology and Oncology, Osaka University Graduate School of Medicine

ABSTRACT

Hematopoietic stem cells (HSCs) are characterized by two distinct abilities, that is, self-renewal ability and multipotency. To keep homeostasis of hematopoiesis and protect exhaustion of HSCs throughout the life, most of HSCs are kept quiescent and only a limited number of HSCs enter cell cycle to supply mature blood cells. Cell cycle state of HSCs is crucially regulated by external factors such as cytokines, adhesion molecules, Notch ligands, and Wnt signals in the bone marrow (BM) microenvironment, so called hematopoietic niche. In addition, intrinsic molecules expressed in HSCs such as transcription factors and cell cycle regulatory molecules also control their growth and differentiation. To utilize HSCs more efficiently and to develop new therapeutic strategies for various diseases, it is of particular interest to expand HSCs *ex vivo*. At present, three clinical studies, in which cord blood HSCs were *ex vivo* expanded by cytokines and transplanted into patients with hematologic malignancies, have been performed. However, the expanded HSCs did not shorten the recovery of hematopoiesis.

[*] **Address correspondence to** Hirokazu Tanaka, M.D., Ph.D. Department of Regenerative Medicine, Institute of Biomedical Research and Innovation, Kobe, Japan. Address: 2-2 Minatojima-Minamimachi, Chuo-ku, Kobe, 650-0047 Japan. TEL: +81-78-304-5773. FAX: +81-78-304-5774. E-mail: htanaka@fbri.org

So, further novel strategies to expand HSCs more efficiently and to fasten hematopoiesis from HSCs are required by modifying the function of the molecules that regulate self-renewal of HSCs.

INTRODUCTION

HSCs are characterized by two distinct abilities; self-renewal ability and multipotency. With these activities, HSCs are capable of maintaining a life-long supply of all lineages of hematopoietic cells according to systemic needs. The durability of the output potential of HSCs is believed to be dependent on their ability to execute self-renewal divisions; that is, an ability to proliferate without activation of a latent readiness to differentiate along restricted lineages. *In vivo*, to maintain homeostasis of hematopoiesis and protect exhaustion of HSC population, most of HSCs are kept quiescent and only a limited number of cells enter cell cycle to supply mature blood cells. During this cell division, HSCs are obliged to undergo self-renewal, differentiation, or apoptosis. This step is controlled by external stimuli transmitted from the bone marrow (BM) microenvironment, including cytokines signals, adhesion molecules, Notch ligands, and Wntsignals. Also, intrinsic factors expressed in HSCs, such as transcription regulators and cell cycle regulatory molecules, are crucially involved in this regulation.

Human umbilical cord blood (CB) is a useful source of HSCs for transplantation. In fact, during the last few years an increasing number of patients have received CB transplants [1]. However, its clinical application is restricted because of the insufficient number of HSCs in each CB sample for most of adult patients. Also, compared with transplantation using HSCs from BM or mobilized into peripheral blood, the recovery of hematopoiesis is rather delayed in CB transplantation, partly due to the insufficient number of transplanted HSCs/hematopoietic progenitor cells (HPCs) and to the persistent quiescence of CB HSCs, which is often accompanied by lethal complications [1]. Therefore, it is of particular interest to expand CB HSC/HPCs *ex vivo* and to develop strategies for hastening hematopoietic recovery after CB transplantation *in vivo* [2-4].

In this paper, we will review recent papers describing molecular mechanisms governing the stemness of HSCs. Also, by referring the results of three clinical trials, in which *ex vivo* expanded cord blood HSCs were transplanted, we will discuss how we can solve the problems observed in these studies by modulating the function of molecules that regulate self-renewal of HSCs.

CHARACTERISTIC OF HSCS

The procedure for the purification of HSCs has made great progress along with the identification of molecular markers that characterize the cells having reconstitution activities in transplanted mice. The most primitive murine HSCs are considered to be with the $CD34^{low/-}$c-Kit^+Sca-1^+Lin$^-$ (CD34$^-$KSL) phenotype, since a single cell with this phenotype could reconstitute whole hematopoiesis *in vivo* with high probability [5]. In addition to the specific surface phenotype, HSCs present in steady-state adult mouse BM are functionally characterized by their ability to excrete Rhodamine-123 and Hoechst 33342 [6, 7]. When adult mouse BM cells are stained with Hoechst 33342, exposed to the UV light, and examined at 2 emission wavelengths simultaneously, HSCs are found in the rare side population (SP) with the dim fluorescence because of this ability [8, 9]. The low fluorescence of HSCs after staining with Rhodamine-123 and Hoechst 33342 is attributed to their selective expression of different ABC transporters, P-glycoprotein and bcrp-1, respectively [10,11].

In addition, the cells having the strongest dye efflux capacity (Tip-SP cells) with the CD34-KSL phenotype were shown to be the most primitive HSCs, which can reconstitute long-term hematopoiesis with almost 100% probability even after the single cell transplantation [12]. The cells in the SP fraction is considered to be in G0 phase, and this state is supposed to be regulated by "hematopoietic niche" in the BM as described later. On the other hand, in steady-state human BM, a majority of HSCs having long-term reconstitution activity express CD34 [13], and the most primitive human HSCs are considered to exist in the population with Lineage (CD3, CD4, CD8, CD11b, CD19, CD20, hCD56, and glycophorin A)$^-$CD34$^+$CD38$^-$ phenotype [14]. Whereas CD34$^+$ has been utilized as a marker of HSCs, recent reports indicated that SRCs (SCID-repopulating cells) are more concentrated in CD133$^+$ cells than in CD34$^+$ cells [15,16]. That is, Suzuki et al. demonstrated that CB CD133-sorted cells contained an approximately 4.5-fold greater absolute number of SRCs than CD34-sorted cells [17]. Further clinical studies comparing CD34$^+$ and CD133$^+$ cells would determine which is a better phenotype to collect and evaluate human HSCs.

Cytokines Involved in Stemness Regulation in HSCs

A number of cytokines regulate growth, differentiation, and survival of HSCs both positively and negatively. Among these, stem cell factor (SCF), Flt3 ligand (FL), thrombopoietin (TPO), interleukin-3 (IL-3), and IL-6 are known to promote the growth of HSCs *in vitro* [18-20]. In fact, Sl/Sl and W/W mice each having homozygous defect in the SCF gene and its receptor c-*kit* gene reveal severe anemia [21]. Also, total number of HSCs was reduced in the BM of c-*mpl* (TPO receptor)-null mice [22]. In addition, c-*mpl*$^{-/-}$ HSCs revealed severely decreased activities in reconstitution assays. These lines of evidence indicate that cytokine signals are required for the growth and survival of HSCs *in vivo* as well as *ex vivo* [23].

TGF-β1 is a 25 kd protein produced by stromal cells and hematopoietic progenitors, which induces the growth arrest in HSCs in autocrine and/or paracrine manners [24-28]. Using antisense oligonucleotides, it was demonstrated that the inhibition of TGF-β1 production could release HSCs in the umbilical CB or BM from the quiescent state [29-32]. Furthermore, the inhibition of the TGF-β1 signaling pathways in human HSCs using blocking antibodies against TGF-β1 or its receptor allowed quiescent cells to enter cell cycle [33]. TGF-β1 had been supposed to induce cell-cycle arrest in various cell types including HSCs through cyclin-dependent kinase inhibitors (CKIs), p21^{WAF1} (p21) and p27^{Kip1} (p27) [34-40]. However, a recent paper provided evidence that TGF-β1 induced growth arrest independently of p21^{WAF1} or p27^{Kip1} using HSCs and progenitor cells lacking both p21^{WAF1} and p27^{Kip1} [41]. As for the other possible mechanism of TGF-β1-induced growth arrest, TGF-β1 was reported to induce the expression of the other CKI, p15I^{NK4B} (p15), [42, 43] and downregulate the expression of c-Kit, FLT3, and IL-6 receptor on HSCs, thereby disrupting cytokine-dependent growth signals [44, 45].

In contrast, another TGF-β super family protein, bone morphogenetic protein-4 (BMP-4) was reported to induce self-renewal of HSCs [46].

At present, the utilization of cytokines is the most promising and practical strategy for the *ex vivo* expansion of HSCs. To establish the culture conditions most suitable for expansion of HSCs, a number of investigators have employed various cytokine combinations [47, 48]. When their effects were compared by long-term reconstitution assays in transplanted mice, the combination of SCF, FL, TPO, and IL-6/soluble IL-6 receptor (sIL-6R) was found to expand HSCs most efficiently, with a 4.2-fold increase in SCID-repopulating cells (SRC) [49]. Several patients were already transplanted with cytokine-expanded CB HSCs without serious toxicities as described later [136-138]. However, cytokine-expanded CB HSCs did not shorten the nadir period after transplantation, indicating the limited usefulness of cytokines for *ex vivo* expansion of CB HSCs. Thus, further improvement is necessary to prepare more efficient HSCs [50].

Effects of the BM Microenvironment "Hematopoietic Niche" on Stemness of HSCs

As were the cases with gut and certain skin stem cells [51, 52], HSCs receive critical signals for proliferation and differentiation from the BM microenvironment called "hematopoietic niche" (Figure 1), which consists of stromal cells and the extracellular matrix (ECM) [53-55]. ECM is composed of a variety of molecules such as fibronectin (FN), collagens, laminin, and proteoglycans. ECM in the BM is not merely an inert framework but mediates specialized functions [56-60]. Some components of ECM bind to growth factors produced by stromal cells and immobilize them around cells, which gives spaces where hematopoietic cells and growth factors colocalize. In addition, ECM can bind to glycoproteins expresssed on HSCs. FN, collagens, and laminin are ligands for integrins that not only control anchorage, spreading, and migration of HSCs but also activate signal transduction pathways in these cells [56-58, 61].

Two groups individually generated mice lacking the BMP receptor type A (BMPRIA) and those engineered to produce osteoblast-specific, activated parathyroid hormone (PTH) and PTH-related protein (PTHrP) receptors (PPRs) [62,63]. In these mice, the osteoblast population was found to increase in the specific regions of bone, "trabecular bone-like areas''. Also, the increase of the osteoblast population caused the parallel increase of the HSC population, particularly long-term repopulating HSCs. As for this mechanism, Zhang et al. demonstrated that the long-term HSCs were attached to spindle-shaped N-cadherin^{+}CD45^{-} osteoblastic (SNO) cells. Two adherent junction molecules, N-cadherin and β-catenin, were asymmetrically localized between the SNO cells and the long-term HSCs, suggesting that SNO cells function as a key component of the niche to support HSCs, and that BMP signaling through BMPRIA controls the number of HSCs by regulating niche size. Meanwhile, in the latter study, Calvi et al. demonstrated that PPR-stimulated osteoblasts produced high levels of the Notch1 ligand, Jagged1, and supported the activity of HSCs through the Notch signaling.

Together, these papers indicate that the interaction with osteoblasts contributes to the maintenance of HSCs.

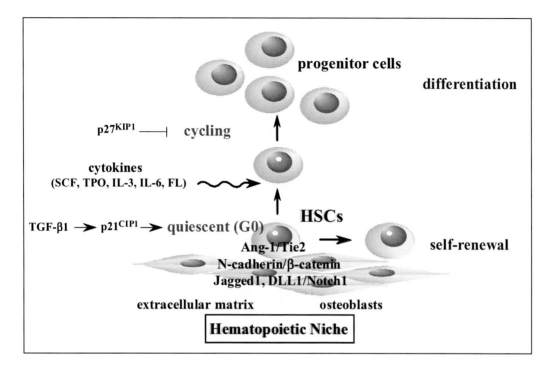

Figure 1. Effects of BM Microenvironment on Cell Cycle of HSCs.

HSCs expressing the receptor tyrosine kinases, Ties, were quiescent, and the ligand for Tie2, Ang-1, which is expressed on endothelial cells and HSCs, enhanced the quiescence of HSCs and their adhesion to fibronectin and collagen [64,65]. Therefore, it was assumed that the Ang-1/Tie2 signaling pathway plays some role to keep HSCs in quiescent. In accord with this hypothesis, a recent paper proved that Tie2[+] HSCs were in close contact with sub-endoesteal osteoblasts expressing Ang-1, and that these Tie2[+] cells were included in SP and in G0 phase of cell cycle [66]. These results suggest that HSCs attaching to the specific osteoblasts in the hematopoietic niche are kept quiescent and protected from the myelosuppressive stress such as the treatment with 5-Fluorouracil (5-FU), a cell cycle-specific myelotoxic agent that kills cycling cells. However, it remains unknown which fraction of osteoblasts expresses Ang-1 and how it is regulated. Furthermore, the molecular mechanisms through which Tie2/Ang-1 signaling prevents cell cycle entry also remain elusive.

Effects of Notch ligand, Wnt, and Sonic Hedgehog (Shh) Signals on Self-Renewal of HSCs

Besides cytokines and molecules consisting of the extracellular matrix, various stimuli such as the Notch ligand, Wnt, and Shh are transmitted to HSCs in the BM microenvironment. The activation of Notch transmembrane receptors expressed on HSCs by their ligand (Delta or Jagged) expressed on stromal cells promotes self-renewal of HSCs [67-71]. Karanu et al. reported that a soluble form of Jagged-1 can enhance the expansion of human CD34$^+$ HSCs when added to liquid cultures with SCF, FL, IL-6, IL-3, and G-CSF, indicating the potential usefulness of soluble Jagged-1 for promoting *ex vivo* expansion of HSCs. It was also reported that a soluble form of Delta-like1 augmented cytokine-dependent *ex vivo* expansion of HSCs in CB CD133$^+$ cells measured by SRCs [17]. However, since Notch-1 has propensity to induce lymphoid differentiation rather than myeloid differentiation, further studies are required to verify the usefulness and to establish the utility of these approaches. Nonetheless, these strategies would be a promising method to expand HSCs *ex vivo*. As for the critical target molecule of Notch signals that mediates self-renewal of HSCs, we found that c-Myc was transcriptionally induced by Notch [72]. In addition, the ectopic expression of c-Myc induced the growth of HSCs without disrupting their biologic properties in terms of surface phenotypes, colony-forming abilities, and reconstituting abilities. Thus, c-Myc was supposed to play a major role in self-renewal of HSCs as an effector molecule of Notch signals.

Like Jagged1/Notch, a number of Wnt proteins are expressed in the BM and their receptor frizzled was detectable on BM-derived HSC/HPCs [73,74]. In the absence of Wnt-mediated signaling, β-catenin is degraded by the ubiquitin/proteosome pathway. Wnt signaling through frizzeled inhibits the degradation of β-catenin, resulting in the accumulation of β-catenin associating with T-cell factor (TCF)/lymphoid-enhancer-binding factor (LEF)-family transcription factors, and these proteins regulate the transcription of downstream target genes. As for the effects of Wnt on HSCs, purified Wnt3a was shown to expand HSCs isolated from Bcl2-transgenic mice *ex vivo* [75]. In addition to Wnt3a that activates the canonical pathway through Frizzled/β-catenin/TCF/LEF, non-canonical Wnt, Wnt-5a, was also reported to expand HSCs *in vitro* [76]. However, its mechanisms remain to be clarified. Also, retrovirally expressed a constitutively active form of β-catenin enhanced proliferation of a phenotypically defined murine HSC population [77]. Limiting dilution assays indicated that the induction of activated β-catenin led to over 50-fold increase in HSC numbers after 1-week culture. As for the mechanism of Wnt-mediated proliferation of HSCs, it was demonstrated that the activation of Wnt signaling induced the increased expression of HoxB4 and Notch1 in HSCs,. Glycogen synthase kinase-3(GSK-3) is a constitutively active serine-threonine kinase, which form the destruction complex with Axin and adenomatous polyposis coli (APC). Association with this complex leads to ubiquitilation of β-catenin and subsequent proteosomal degradation. Recently, Trowbridge et al. reported beneficial effects of post transplantation treatment with an ATP-competitive GSK-3 inhibitor on human HSC engraftment [78]. These reports suggested that Wnt signaling is important for the *in vitro* and *in vivo* self-renewal of HSCs. However, it is also reported that constitutive activation of β-catenin enforced cell cycle entry of HSCs, thereby, exhausting the long-term repopulating cell pool and leading to hematopoietic failure associated with loss of multilineage differentiation [79,80]. Therefore, fine-tuned (in terms of expression level and duration) Wnt stimulation is required for normal hematopoiesis and critical for therapeutic HSC expansion.

Shh is a family member of human homologs of *Drosophila* Hedgehog (Hh) and expressed on the cell surface as transmembrane proteins. Hh signals can be mediated through cell-to-cell contact between adjacent cells expressing the Patched (Ptc) receptor. Alternatively, NH2-terminal cleavage of Hh can generate a soluble Hh ligand that can interact with distal cells expressing Ptc [81,82]. In the BM, Shh and their receptors Ptc and Smoothened (Smo) are expressed in highly purified HSCs. Cytokine-induced proliferation of HSCs was inhibited by the anti-Hh Ab, implying that endogenously produced Hh proteins play a role in the expansion of HSCs. Conversely, the addition of soluble forms of Shh increased the number of HSCs with pluripotent repopulating abilities. In addition, Noggin, a potent BMP-4 inhibitor, was found to inhibit the mitogenic effects of Shh, indicating that Shh signaling acts upstream of BMP-4 signaling in the proliferation of HSCs [83].

Transcription Factors That Regulate the Stemness of HSCs

In addition to extrinsic factors, accumulated evidence indicates that the stemness of HSCs is regulated by intrinsic transcription regulatory factors, such c-Myc, c-Myb, GATA-2, HOX proteins, and Bmi-1 (Figure 2).

Willson et al. provided genetic evidence for function of c-Myc in the homeostasis of HSCs [84]. c-Myc-deficient HSCs bound to the BM niche too tightly and revealed impaired differentiation, which was correlated with the up-regulation of N-cadherin and a number of adhesion receptors, suggesting that c-Myc was required for the release of HSCs from the stem cell niche. In accord with this finding, endogenous c-Myc is differentially expressed and induced upon differentiation in long-term HSCs. Thus, c-Myc was assumed to control the balance between stem cell self-renewal and differentiation, presumably by regulating the interaction between HSCs and their niche.

A transcriptional factor, c-Myb promotes the growth of HSCs, probably through the induction of c-*myc* and upregulated expression of c-*kit* and Flt3 [85,86], and c-Myb-deficient mice die at embryonic day 15.5 (E15.5) due to the defect of definitive hematopoiesis [87]. Similarly, GATA-2$^{-/-}$ mice are embryonic lethal around E11.5 because of the defect in the development and/or maintenance of HSCs [88]. However, since functional roles of GATA-2 in the growth of HSCs are still controversial [89-92], it remains unknown whether GATA-2 by itself enhances or suppresses the growth of HSCs.

Among Hox family of transcription factors, HOXB4 is of particular remark during the last few years because its retroviral gene transfer induced ~40-fold murine and ~30-fold human HSC expansion *ex vivo,* suggesting its usefulness in clinical application [93-95]. Regarding its clinical use, it was initially concerned that constitutive expression of HOXB4 in HSCs might cause leukemia. This is because deregulated expression of HOXB8 was found in myeloid leukemia, and HOX family genes are sometimes involved in leukemogenic chromosomal translocations such as t(7;11)(p15;p15) yielding NUP98-HOXA9 [96,97]. However, HSCs engineered to express HOXB4 reconstituted all hematopoietic lineages in transplanted mice without causing leukemia, indicating that HSCs overexpressing HOXB4 were still under the control of hematopoietic system *in vivo* [94]. However, very recently, HOXB4 overexpression in collaboration with insertional mutagenesis by virus integration was reported to induce myelomonocytic leukemia in the canine model [98]. Therefore, before clinical application, it is crucial to establish the safe measures as to the retroviral gene transfer strategy. To eliminate any deleterious effects caused by stable HOXB4 gene transfer, Krosl et al. tried to expand murine HSCs by delivering HOXB4 protein [99]. In this study, cell membrane-permeable, recombinant TAT-HOXB4 protein was added to the culture medium, resulting in a five-fold net expansion of HSCs. Although TAT-HOXB4 was delivered with high efficiency, its half-life was estimated as only 1 h. Meanwhile, Amsellem et al. tried to expand human CB HSCs using HOXB4 protein, which was secreted into the culture supernatant from cocultured MS-5 murine stromal cells [100]. This approach increased SRCs 2.5-fold. However, the efficiency of protein delivery was not so high, and the coculture system would not be practical for clinical applications.

HOX proteins interact with the non-HOX homeobox protein, PBX1, and regulate the expression of target genes both positively and negatively,. We recently synthesized a decoy peptide containing the YPWM motif from HOX proteins, which was predicted to act as a HOX mimetic, and analyzed its effects on self-renewal of human cord blood CD34$^+$ cells [101](Figure 3).

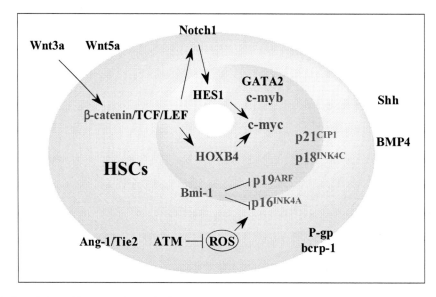

Figure 2. Regulation of Stemness by Intrinsic Factors in HSCs.

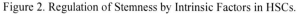

This decoy peptide was delivered into about 70% of CD34$^+$ cells. By examining the expression of HOX target genes c-*myc* and *p21*$^{wafl/cip1}$, we confirmed that decoy peptide enhanced HOX functions. After 7-day culture in serum-free medium containing a cytokine cocktail, the decoy peptide increased numbers of CD34$^+$ cells and primitive multipotent progenitor cells (CFU-Mix) approximately two-fold compared to control cultures. Furthermore, CD34$^+$ cells treated with the decoy peptide reconstituted hematopoiesis in NOD/SCID mice more rapidly and more effectively than control cells (more than two-fold greater efficiency as determined by a limiting dilution method). In addition, decoy peptide-treated CD34$^+$ cells were able to repopulate secondary recipients. Therefore, the decoy peptide will be a promising novel tool for the safe *ex vivo* expansion of human HSC/HPCs in combination with growth factors and/or other approaches.

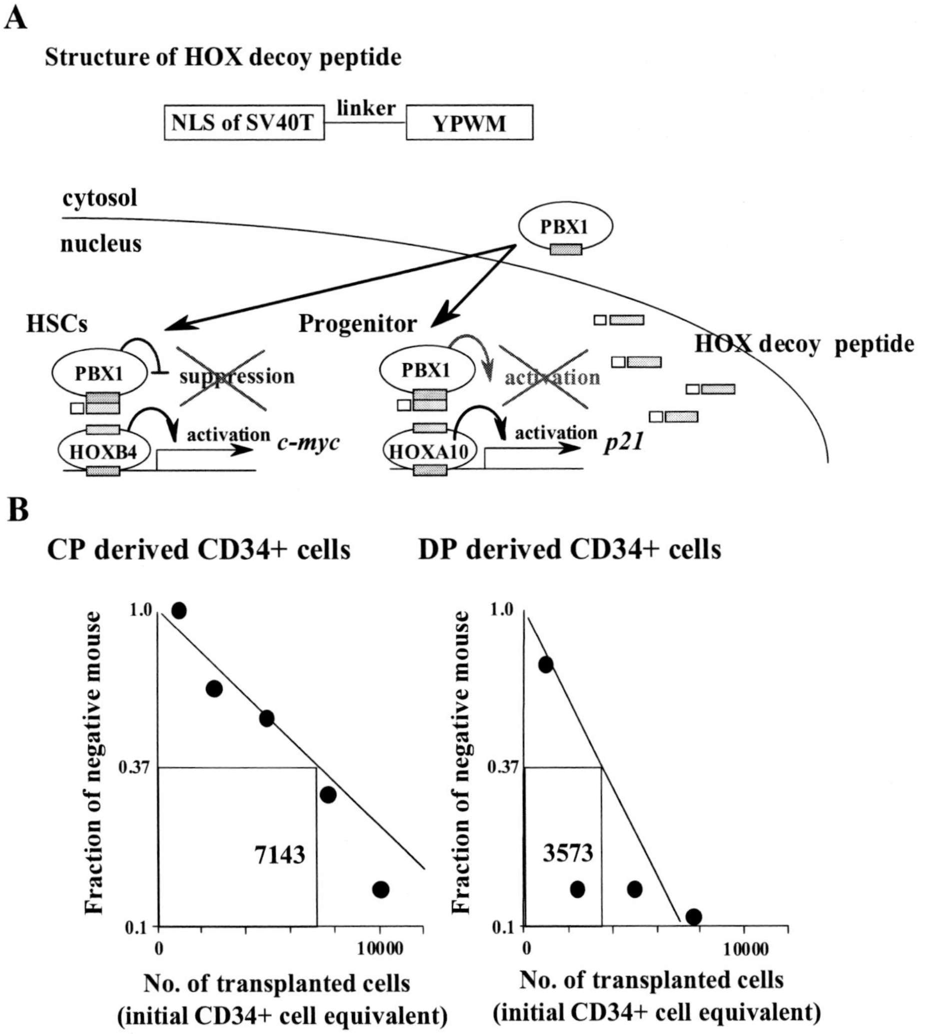

A

Structure of HOX decoy peptide

B

CP derived CD34+ cells   **DP derived CD34+ cells**

No. of transplanted cells   No. of transplanted cells
(initial CD34+ cell equivalent)   (initial CD34+ cell equivalent)

Figure 3. Peptide minetics and *ex vivo* expansion of human HSC/HPCs. A. The HOX decoy peptide contains the YPWM motif of HOX, utilized for the cooperative interaction with PBX1, and the nuclear localization signal (NLS) of the SV40 large T antigen. PBX1 negatively regulates HOXB4-mediated c-*myc* transcription in HSCs, while it promotes HOXA10-mediated p21$^{waf1/cip1}$ expression in myelomonocytic progenitors. Hox decoy peptide was supposed to inhibit both positive and negative effects of PBX1 on HOX-mediated transcription. B. The frequency of LTR-HSCs was calculated to be 1/7143 control peptide (CP)-treated CB CD34^{+} cells 6 weeks after the transplantation to NOD/SCID mice. In contrast, the frequency of LTR-HSCs in HOX decoy peptide (DP)-treated cells was calculated to be 1/3573. Accordingly, the expansion of LTR-HSCs by the decoy peptide was estimated as 2.0-fold.

In addition to HOXB4, other HOX homeobox transcription factors play important roles in the proliferation and differentiation of hematopoietic cells [102,103], however, their physiological functions and roles in leukemogenesis have not been elucidated. Bmi-1, a member of the Polycomb Group family of transcriptional repressors [104], was recently shown to be essential for maintenance of adult self-renewing HSCs [105]. Although the number of HSCs in the fetal liver of Bmi-1$^{-/-}$ mice was normal, the number of HSCs was markedly reduced in postnatal Bmi-1$^{-/-}$ mice. Furthermore, transplanted fetal liver and bone marrow cells obtained from Bmi-1$^{-/-}$ mice were able to contribute to hematopoiesis only transiently. Regarding this mechanism, in accord with the previous data obtained from embryonic fibroblasts [106], the microarray analysis on the BM mononuclear cells showed that the expression of p16^{INK4A} (p16) and p19^{ARF} was upregulated in Bmi-1$^{-/-}$ BM cells. On the other hand, it was reported that the increased expression of Bmi-1 promotes HSC self-renewal in mouse studies [107]. Furthermore, very recently, Rizo, et al. reported that overexpression of Bmi-1 induced *ex vivo* expansion for over 4 month of human CB CD34$^+$ cells in liquid cultures without loosing the potential of stem/progenitor cells [108]. Their data indicate that Bmi-1 is one of the important modulator of human HSCs self-renewal and suggest that it can be a potential target for therapeutic manipulation of human HSCs.

Roles of Cell Cycle Inhibitors, p21 and p27, in HSCs and Progenitor Cells

During the last decade, a number of cell cycle regulatory molecules such as cyclins, cyclin-dependent kinases (CDKs) and CKIs have been identified and their roles and regulation have been well characterized in various types of cells [109-111].

Cell cycle is positively regulated by CDKs associated with cyclins, and their activities are negatively regulated by CKIs also included in these complexes at the same time. CKIs are classified into two families based on their structures and CDK targets. One class of inhibitors including p21, p27, and p57 ^{KIP2} share a CDK2-binding motif in the N-terminus and inhibit the activities of cyclinD-, E-, and A-dependent kinases. The other class of inhibitors also known as the INK4 family, including p16, p15, p18^{INK4C} (p18), and p19 ^{INK4D}, contain fourfold ankyrin repeats and specifically inhibit CDK4 and CDK6. Members of both families are important for executing cell cycle arrest in response to a variety of stimuli such as DNA damage, contact inhibition, and TGF-β1 treatment. Embryonic fibroblasts obtained from p21$^{-/-}$ mice had a defect in their ability to achieve cell cycle arrest after irradiation [112,113], and antisense oligonucleotides against p21 was shown to release human mesenchymal cells from G0 phase [114]. As for the roles for p21 in hematopoiesis, the expression level of p21 was initially reported to be low in CD34$^+$ cells [115,116], and p21$^{-/-}$ mice did not exhibit an apparent hematologic defect [112,113]. However, in a subsequent analysis, Cheng *et al* found that p21 was highly expressed in the quiescent stem cell-like fraction of BM cells [117]. They also found that, under normal homeostatic conditions, the proportion of quiescent HSCs in G0 phase was reduced and that total number of HSCs increased in p21$^{-/-}$ mice. In accord with these findings, when p21$^{-/-}$ mice were treated with 5-FU, the survival percentage was much lower in p21$^{-/-}$ mice than in littermate controls. They also directly assessed self-renewal ability of HSCs using a serial transplantation approach. As a result, no mice transplanted with p21$^{-/-}$ BM cells survived after the fifth transplant due to the exhaustion of HSC population, whereas those transplanted with p21$^{+/+}$ BM cells had a 50% survival. Together, these results indicate that p21 is a key molecule that restricts cell cycle entry of HSCs, thereby keeping their pool and preventing their exhaustion under certain stress conditions. In consistent with these results, Stier *et al.* reported that p21-antisense transduced by the lentivirus vector released human CD34$^+$CD38$^-$ cells from the quiescent state and induced an approximately 2~3-fold expansion of SRCs without loosing multipotency [118], However, further research is required to determine whether human HSCs lacking p21 have the long-term reconstitution abilities.

p27 is molecularly distinct from p21 in its carboxyl terminus; it interacts with similar, though not identical, cyclin-CDK complex and lacks p53-regulated expression. In hematopoietic system, the expression of p27 is observed in more mature progenitors than that of p21 [115,116]. Mice homozygously lacking p27 have a larger body and hyperplasia of most organs including hematopoietic organs [119-121]. In striking contrast to p21$^{-/-}$ mice, the number, cell cycling and self-renewal of HSCs were normal in p27$^{-/-}$ mice, while these mice had an increase in hematopoietic progenitor cells [122]. In addition, these progenitor cells in p27$^{-/-}$ mice were more proliferative than p27$^{+/+}$ progenitor cells. Furthermore, progenitor cells from p27$^{-/-}$ mice were able to expand and regenerate hematopoiesis after serial transplantation, while p27$^{+/+}$ progenitors were markedly depleted. Thus, p21 and p27 govern stem and progenitor cell populations divergently.

Roles for the INK4 Family in Self-Renewal of HSCs and as Tumor Suppressor Genes

In addition to p21 and p27, the INK4 family of CKIs is also implicated in the regulation of HSCs numbers and self-renewal. Mice deficient for p18 had an increased number of HSCs in the bone marrow. Also, competitive repopulation assays showed that p18$^{-/-}$ HSCs were far more competitive than normal HSCs with 14-fold activities. In contrast to p21$^{-/-}$ HSCs, the exhaustion of p18$^{-/-}$ HSCs was not observed during serial bone marrow transplants, indicating that p18 is a strong inhibitor limiting the potential of stem cell self-renewal *in vivo* [123]. From this result, it was speculated that downregulation of the expression of p18 in HSCs, using antisense oligonucleotide or small interfering RNA (siRNA), would be useful for the *ex vivo* expansion of HSCs.

p16 is highly expressed in CD34$^+$ cells, and its expression is downregulated during differentiation process towards all lineages [124]. Nonetheless, since p16$^{-/-}$ mice did not show an apparent abnormality in hematopoiesis, p16 was supposed to be dispensable for the quiescence of HSCs [125,126]. However, Ito et al. found that p16 was essential for reconstitution activities of HSCs but not for proliferation or differentiation of progenitors from the analysis of mice lacking one of the cell cycle check point kinase, "ataxia telangiectasia mutated" (Atm) [127]. In this analysis, Atm$^{-/-}$ mice older than 24 weeks developed progressive bone marrow failure due to a defect of HSC function, which was associated with elevated reactive oxygen species (ROS). Furthermore, they proved p16-retinoblastoma (Rb) pathway activated by ROS was critical for the defective function of HSCs. From these results, they concluded that the self-renewal ability of HSCs depend on ATM-mediated inhibition of oxidative stress and p16-RB pathway. Furthermore, Janzen et al. recently demonstrated that ageing causes an increase in p16 and intrinsic p16 suppress the proliferation of HSC/HPCs in the bone marrow. And they supposed that inhibition of p16 may ameliorate the physiological impact of ageing on stem cells and thereby improve injury repair in aged tissue [128].

In contrast to the expression pattern of p16, the expression of p15 is not detected in CD34$^+$ cells, but increased specfically during myeloid differentiation [124,129]. However, the functional role of p15 in HSCs remained to be clarified. Both p16 and p15 inhibit the function of cyclin D-CDK4/6 complex and suppress the phosphorylation of pRb, thereby inducing cell cycle arrest at G0/G1 phase. Especially, under tumorgenetic stress such as the presence of oncogenic ras gene, p16 and p15 are induced to express and suppress tumor progression through the induction of premature senescence [130,131]. With these activities, both p16 and p15 are supposed to act as tumor suppressor genes. In fact, inactivation and/or deletion of p16 and p15 genes are observed in various human cancers very frequently [132,133]. As for hematological malignancies, their defects caused by the homozygous deletion or methylation were observed in a substantial proportion of AML, ALL, and myelodysplastic syndromes (MDS) cases [134-136]. These results indicate that appropriate cell cycle control, particularly at the stage of stem/progenitor cells, is required for maintaining normal hematopoiesis, and have to be uppermost when manipulating HSCs *ex vivo*.

PILOT TRIALS OF TRANSPLANTATION WITH *EX VIVO* EXPANDED CB CELLS

To date, three groups of investigators have utilized *ex vivo*-amplified CB HSCs for transplantation (Table 1). Shpall *et al*. isolated CD34$^+$ cells from CB. Then, forty percent of the isolated cells were expanded in medium containing SCF, G-CSF, and TPO for 10 days, and the remaining 60% were immediately transplanted or stored frozen until transplantation [137]. After high-dose chemotherapy, 37 patients (25 adults, 12 children) were transplanted with expanded CD34$^+$ cells and non-expanded cells with a median dose of 0.99×10^7 nucleated cells per kilogram. The median time to engraftment of neutrophils (neutrophil count $>500/\mu l$) was 28 days (range 15-49 days) and that of platelets (platelet $>20,000/\mu l$) was 106 days (range 38-345 days). From this study, the authors concluded that, although the transplantation with *ex vivo* expanded CB cells was feasible and safe, expanded HSCs did not improve the time to engraftment in recipients. In a phase I trial, Jaroscak *et al*. transplanted CB HSCs expanded by PIXY321, FL, and EPO into 28 patients with a median dose of 2.4×10^7 nucleated cells per kilogram [138]. They also concluded that the *ex vivo*-expanded CB HSCs were not effective in shortening the recovery period, probably due to insufficient expansion of CD34$^+$ cells in their culture system. In contrast, Shpall *et al*. expanded CD133$^+$ cells isolated from CB in liquid culture containing with SCF, IL-6, FL, and polyamine copper chelator, TEPA for 21 days, which led to the expansion of a cell population that displays phenotypic and functional characteristics of HSC/HPCs [139]. As a result of transplantation into the 10 patients, the median time to engraftment was 27 days for neutrophils (range 16-46 days) and 48 days for platelets (range 27-96 days). This preliminary result showed that a shorter time to neutrophil engraftment was correlated with total TNC per kilogram infused, and a trend with CD34$^+$ cells per kilogram infused. These pilot studies indicated that a novel strategy, which can expand HSCs more efficiently without losing their functions and properties, is absolutely prerequisite for the useful clinical application of HSC/HPCs expansion.

CONCLUSION

Owing to the recent advance in stem cell biology, various molecules involved in self-renewal of HSCs have been identified. Also, it has been clarified how the activities of these molecules are regulated. So, the establishment of a novel useful strategy for *ex vivo* expansion of HSCs and its clinical application will be realized in the near future (Figure 4), However, further studies are required to disclose the whole feature of stemness regulation in HSCs. These studies would undoubtfully enable us to utilize HSCs more efficiently.

Table 1. Pilot trials of UCB transplantation with *ex vivo* expanded CB cells

Group	Shpall et al. [136]	Jaroscak et al. [137]	Shpall et al. [138]
Cytokine cocktail	SCF+G-CSF +TPO	GM-CSF+IL-3 +FL+EPO	SCF+IL-6+FL +TPO+TEP A
Culture duration	10 days	12 days	21 days
Fold expansion ofTNCs*	56 (0.1-278)	2.05 (0.06-10.19)	207 (2-616)
Fold expansion of CD34*	4 (0.1-20)	0.5 (0.09-2.45)	N.A.
Infused TNCs($\times 10^7$/kg)*	0.99 (0.28-8.5)	2.4 (1.0-8.5)	1.8 (1.1-6.1)
Infused CD34($\times 10^5$/kg)*	1.04 (0.97-31.1)	0.22 (0.001-2.59)	1.6 (0.4-49.9)
No. of enrolled patients	37	2 8	10
Engraftment			
Neutro>500*	Day 28 (15-49)	Day 22 (13-40)	Day 27 (16-46)
Plt>50,000*	Day106 (28-345)	Day 94 (41-370)	Day 48 (27-96)
Graft failure	0/30	3/24	0/7
acute GVHD (>II)	40%	36%	43%

* median (range).
TNCs: total nuclear cells.
N.A.: no assessment.

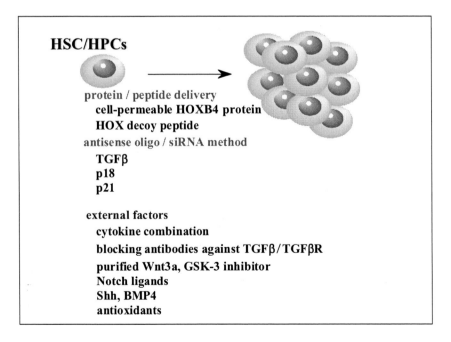

Figure 4. Strategies of therapeutic HSC/HPCs expansion.

ACKNOWLEDGEMENT

This paper was supported in part by grand aid from Sankyo Foundation of Life Science.

REFERENCES

[1]   Benito, AI; Diaz, MA; Gonzalez-Vicent, M; Sevilla, J; Madero, L. Hematopoietic stem cell transplantation using umbilical cord blood progenitors: review of current clinical results. *Bone Marrow Transplant.*, 2004, 33, 675-690.

[2]   Devine, SM; Lazarus, HM, Emerson, S.G. Clinical application of hematopoietic progenitor cell expansion: current *status and future prospects. Bone* Marrow Transplant., 2003, 31, 241-252.

[3]   Sauvageau, G; Iscove, NN; Humphries, RK. *In vitro* and *in vivo* expansion of hematopoietic stem cells. *Oncogene.* 2004, 23, 7223-32.

[4]   Sorrentino, BP. Clinical strategies for expansion of haematopoietic stem cells. *Nat. Rev. Immunol.*, 2004, 4, 878-88.

[5]   Osawa, M; Hanada, K; Hamada, H; Nakauchi, H. Long-term lymphohematopoietic reconstitution by a single CD34-low/negative hematopoietic stem cell. *Science,* 1996, 273, 242-245.

[6]   Bertoncello, I; Hodgson, GS; Bradley, TR. Multiparameter analysis of transplantable hemopoietic stem cells, I: the separation and enrichment of stem cells homing to marrow and spleen on the basis of Rhodamine-123 fluorescence. *Exp. Hematol.*, 1985, 13, 999-1006.

[7]   Wolf, NS; Kone, A; Priestley, GV; Bartelmez, SH. *In vivo* and *in vitro* characterization of long-term repopulating primitive hematopoietic cells isolated by sequential Hoechst 33342-rhodamine123 FACS selection. *Exp. Hematol.*, 1993, 21, 614-622.

[8]   Goodell, MA; Brose, K; Paradis, G; Conner, AS; Mulligan, RC. Isolation and functional properties of murine hematopoietic stem cells that are replicating *in vivo. J. Exp. Med.*, 1996, 183, 1797-1806.

[9]   Zhou, S; Schuetz, JD; Bunting, KD; Colapietro, AM; Sampath, J., Morris,JJ; Lagutina, I; Grosveld, GC; Osawa, M. et al. The ABC transporter Bcrp1/ABCG2 is expressed in a wide variety of stem cells and is a molecular determinant of the side-population phenotype. *Nat. Med.*, 2001. 7, 1028-1034.

[10]   Uchida, N; Leung, FYK; Eaves, CJ. Liver and marrow of adult mdr-1α/1β-/- mice show normal generation, function and multi-tissue trafficking of primitive hematopoietic cells. *Exp. Hematol.*, 2002, 30, 862-869.

[11]   Zhou, S; Morris, JJ; Barnes, Y; Lan, L; Schuetz, JD; Sorrentino, BP. Bcrp1 gene expression is required for normal numbers of side population stem cells in mice, and confers relative protection to mitoxantrone in hematopoietic cells *in vivo. Proc. Natl. Acad. Sci. USA.*, 2002, 99, 12339-12344.

[12]   Matsuzaki, Y; Kinjo, K; Mulligan, RC; Okano, H. Unexpectedly efficient homing capacity of purified murine hematopoietic stem cells. *Immunity,* 2004, 20, 87-93.

[13]   Okuno, Y; Iwasaki, H; Huettner, CS; Radomska, HS; Gonzalez, DA; Tenen, DG; Akashi, K. Differential regulation of the human and murine CD34 genes in hematopoietic stem cells. *Proc. Natl. Acad. Sci. USA.* 2002, 99, 6246-51.

[14]   Ishikawa, F; Yasukawa, M; Lyons, B; Yoshida, S; Miyamoto, T; Yoshimoto, G; Watanabe, T; Akashi, K; Shultz, LD; Harada, M. Development of functional human blood and immune systems in NOD/SCID/IL2 receptor {gamma} chain (null) mice. *Blood,* 2005, 106, 1565-73.

[15]   Gordon, PR; Leimig, T; Babarin-Dorner, A; Houston, J; Holladay, M; Mueller, I; Geiger, T; Handgretinger, R. Large-scale isolation of CD133+ progenitor cells from G-CSF mobilized peripheral blood stem cells. Bone Marrow Transplant., 2003, 31, 17-22. [16] de Wynter, EA; Buck, D; Hart, C; Heywood, R; Coutinho, LH; Clayton, A; Rafferty, JA; Burt, D; Guenechea, G; Bueren, JA. et al. CD34+AC133+ cells isolated from cord blood are highly enriched in long-term culture-initiating cells, NOD/SCID-repopulating cells and dendritic cell progenitors. *Stem Cells.* 1998, 16, 387-96.

[16]   Suzuki, T; Yokoyama, Y; Kumano, K; Takanashi, M; Kozuma, S; Takato, T; Nakahata, T; Nishikawa, M; Sakano, S; Kurokawa, M. et al. Highly efficient *ex vivo* expansion of human hematopoietic stem cells using Delta1-Fc chimeric protein. *Stem Cells*, 2006, 24, 2456-2465.

[17]   Nakauchi, H; Sudo, K; Ema. H. Quantitative assessment of the stem cell self-renewal capacity. *Ann. N. Y. Acad. Sci.*, 2001, 93,18-24.

[18]   Petzer, AL; Zandstra, PW; Piret, JM; Eaves.CJ. Differential cytokine effects on primitive (CD34+CD38-) human hematopoietic cells: novel responses to Flt3-ligand and thrombopoietin. *J. Exp. Med.*, 1996, 183, 2551-2558.

[19]   Sitnicka, E; Lin, N; Priestley, GV; Fox, N; Broudy, VC; Wolf, NS; Kaushansky K. The effect of thrombopoietin on the proliferation and differentiation of murine hematopoietic stem cells. *Blood*, 1996, 87, 4998-5005.

[20]   Kitamura, Y; Kasugai, T; Arizono, N; Matsuda, H. Development of mast cells and basophils: processes and regulation mechanisms. *Am. J. Med. Sci.*, 1993, 306, 185-191.

[21]   Alexander, WS; Roberts, AW; Nicola, NA; Li, R; Metcalf, D. Deficiencies in progenitor cells of multiple hematopoietic lineages and defective megakaryocytopoiesis in mice lacking the thrombopoietic receptor c-Mpl. *Blood*, 1996, 87, 2162-2170.

[22]   Kimura, S; Roberts, AW; Metcalf, D; Alexander, WS. Hematopoietic stem cell deficiencies in mice lacking c-Mpl, the receptor for thrombopoietin. *Proc. Natl. Acad. Sci. USA*, 1998, 95, 1195-1200.

[23]   Roberts, AB; Anzano, MA; Wakefield, LM; Roche, NS; Stern, DF; Sporn, MB. Type beta transforming growth factor: a bifunctional regulator of cellular growth. *Proc..Nat.Acad.Sci.USA*, 1985, 82, 119-123.

[24]   Massague, J. The transforming growth factor-beta family. Annual Rev. Cell Biol., 1990, 6, 597-641.

[25]   Eaves, CJ; Cashman, JD; Kay, RJ. Mechanisms that regulate the cell cycle status of very primitive hematopoietic cells in long-term human marrow cultures. II. Analysis of positive and negative regulators produced by stromal cells within the adherent layer. *Blood*, 1991, 78,110-117.

[26]   Nemunaitis, J; Tompkins, CK; Andrews, DF; Singer JW. Transforming growth factor beta expression in human marrow stromal cells. *Eur. J. Hematol.*, 1991, 46,140-145.

[27]   Moore, SC; Theus, SA; Barnett, JB. Bone marrow natural suppressor cells inhibit the growth of myeloid progenitor cells and the synthesis of colony-stimulating factors. *Exp. Hematol.*, 1992, 20, 1178-1183.

[28]   Hatzfeld, J; Li, .L; Brown, EL; Sookdeo, H; Levesque, JP; O'Toole, T; Gurney, C; Clark, SC; Hatzfeld, A. Release of early human hematopoietic progenitors from quiescence by antisense transforming growth factor beta 1 or Rb oligonucleotides. *J. Exp. Med.*, 1991, 174, 925-929.

[29]   Fan, X; Valdimarsdottir, G; Larsson, J; Brun, A; Magnusson, M; Jacobsen, SE; ten Dijke, P; Karlsson, S. Transient disruption of autocrine TGF-beta signaling leads to enhanced survival and proliferation potential in single primitive human hemopoietic progenitor cells. *J. Immunol.*, 2002, 168, 755-762.

[30]   Cardoso, AA; Li, ML; Batard, P; Hatzfeld, A; Brown, EL; Levesque, JP; Sookdeo, H; Panterne, B; Sansilvestri, P; Clark, SC. et al. Release from quiescence of CD34+ CD38- human umbilical cord blood cells reveals their potentiality to engraft adults. *Proc. Natl. Acad. Sci. USA*, 1993, 90, 8707-8711.

[31]   Li, ML; Cardoso, AA; Sansilvestri, P; Hatzfeld, A; Brown, EL; Sookdeo, H; Levesque JP; Clark SC; Hatzfeld J. Additive effects of steel factor and antisense TGF-beta 1 oligodeoxynucleotide on CD34+ hematopoietic progenitor cells. *Leukemia*, 1994, 8, 441-445.

[32]   Fortunel, N; Hatzfeld, J; Kisselev, S; Monier, MN; Ducos, K; Cardoso, A; Batard, P; Hatzfeld, A. Release from quiescence of primitive human hematopoietic stem/progenitor cells by blocking their cell-surface TGF-beta type II receptor in a short-term *in vitro* assay. *Stem Cells*, 2000, 18, 102-111.

[33]   Datto, MB; Li, Y; Panus, JF; Howe, DJ; Xiong, Y; Wang, XF. Transforming growth factor beta induces the cyclin-dependent kinase inhibitor p21 through a p53-independent mechanism. *Proc.Natl. Acad. Sci. USA*, 1995, 92, 5545-5549.

[34]   Landesman, Y; Bringold, F; Milne, DD; Meek, DW. Modifications of p53 protein and accumulation of p21 and gadd45 mRNA in TGF-beta 1 growth inhibited cells. *Cell Signal*, 1997, 9, 291-298.

[35]   Miyazaki, M; Ohashi, R; Tsuji, T; Mihara, K; Gohda, E; Namba, M. Transforming growth factor-beta 1 stimulates or inhibits cell growth via down- or up-regulation of p21/Waf1. *Biochem. Biophys. Res. Commun.*, 1998, 246, 873-880.

[36]   Li, C.Y; Suardet, L; Little, JB. Potential role of WAF1/Cip1/p21 as a mediator of TGF-beta cytoinhibitory effect. *J. Biol. Chem.* 1995, 270, 4971-4974.

[37]   Elbendary, A; Berchuck, A; Davis, P; Havrilesky, L; Bast, RC, Jr; Iglehart, JD; Marks, JR. Transforming growth factor beta 1 can induce CIP1/WAF1 expression independent of the p53 pathway in ovarian cancer cells. *Cell Growth Differ.,* 1994, 5, 1301-1307.

[38]   Ducos, K; Pantern, B; Fortunel, N; Hatzfeld, A; Monier, MN; Hatzfeld, J. p21(cip1) mRNA is controlled by endogenous transforming growth factor-beta1 in quiescent human hematopoietic stem/progenitor cells. *J. Cell Physiol.* 2000, 184, 80-85.

[39]   Fortunel, NO; Hatzfeld, A; Hatzfeld, JA. Transforming growth factor-beta: pleiotropic role in the regulation of hematopoiesis. *Blood,* 2000, 96, 2022-2036.

[40]   Cheng, T; Shen, H; Rodrigues, N; Stier, S; Scadden, DT. Transforming growth factor beta 1 mediates cell-cycle arrest of primitive hematopoietic cells independent of p21(Cip1/Waf1) or p27(Kip1). *Blood,* 2001, 98, 3643-3649.

[41]   Hannon, GJ; Beach, D. p15INK4B is a potential effector of TGF--induced cell cycle arres. *Nature,* 1994, 371, 257-261.

[42]   Li, JM; Nichols, MA; Chandrasekharan, S; Xiong, Y; Wang, XF. Transforming growth factor beta activates the promoter of cyclin-dependent kinase inhibitor p15INK4B through an Sp1 consensus site. *J. Biol. Chem.,* 1995, 270, 26750-26753.

[43]   Sansilvestri, P; Cardoso, AA; Batard, P; Panterne, B; Hatzfeld, A; Lim, B; Levesque, JP; Monier, MN; Hatzfeld, J. Early CD34high cells can be separated into KIThigh cells in which transforming growth factor-beta (TGF-beta) downmodulates c-kit and KITlow cells in which anti-TGF-beta upmodulates c-kit. *Blood,* 1995, 86, 1729-1735.

[44]   Batard, P; Monier, MN; Fortunel, N; Ducos, K; Sansilvestri-Morel, P; Phan, T; Hatzfeld, A; Hatzfeld, JA. TGF-(beta)1 maintains hematopoietic immaturity by a reversible negative control of cell cycle and induces CD34 antigen up-modulation. *J. Cell Sci.,* 2000, 111, 1867-1875.

[45]   Bhatia, M; Bonnet, D; Wu, D; Murdoch, B; Wrana, J; Gallacher, L; Dick, JE. Bone morphogenetic proteins regulate the developmental program of human hematopoietic stem cells. *J. Exp. Med.,* 1999, 189, 1139-48.

[46]   Devine, SM; Lazarus ,HM; Emerson, SG. Clinical application of hematopoietic progenitor cell expansion: current status and future prospects. *Bone Marrow Transplant.,* 2003, 31, 241-252.

[47]   Heike, T; Nakahata, T. *Ex vivo* expansion of hematopoietic stem cells by cytokines. *Biochim. Biophys. Acta.,* 2002, 1592, 313-321.

[48]   Ueda, T; Tsuji, K; Yoshino, H; Ebihara, Y; Yagasaki, H; Hisakawa, H; Mitsui, T; Manabe, A; Tanaka, R; Kobayashi, K. et al. Expansion of human NOD/SCID-repopulating cells by stem cell factor, Flk2/Flt3 ligand, thrombopoietin, IL-6, and soluble IL-6 receptor. *J. Clin. Invest.,* 2000, 105, 1013-1021.

[49]   Fernandez, MN; Regidor, C; Cabrera, R; Garcia-Marco, J., Briz, M; Fores ,R; Sanjuan, I; McWhinnie, A; Querol, S; Garcia, J. et al. Cord blood transplants: early recovery of neutrophils from co-transplanted sibling haploidentical progenitor cells and lack of engraftment of cultured cord blood cells, as ascertained by analysis of DNA polymorphisms. *Bone Marrow Transplant.,* 2001, 28, 355-363.

[50]   Marshman, E; Booth, C; Potten, CS. The intestinal epithelial stem cell. *Bioessays,* 2002, 24, 91-8.

[51]   Nishimura, EK; Jordan, SA; Oshima, H; Yoshida, H; Osawa, M; Moriyama, M; Jackson, IJ; Barrandon, Y; Miyachi, Y; Nishikawa, S. Dominant role of the niche in melanocyte stem-cell fate determination. *Nature,* 2002, 416, 854-60.

[52]   Zuckerman, S; Wicha, MS. Extracellular matrix production by the adherent cells of long-term murine bone marrow cultures. *Blood,* 1983, 61, 540-547.

[53]   Campbell, A; Wicha, MS; Long, M. Extracellular matrix promotes the growth and differentiation of murine hematopoietic cells *in vitro. J. Clin. Invest.,* 1985, 75, 2085-2090.

[54] Long, MW; Briddell, R; Walter, AW; Bruno, E; Hoffman, R., Human hematopoietic stem cell adherence to cytokines and matrix molecules. *J. Clin. Invest.*, 1992, 90, 251-255.

[55] Adams, JC; Watt, FM. Regulation of development and differentiation by the extracellular matrix. *Development*, 1993, 117, 1183-1198.

[56] Long, M.W. Blood cell cytoadhesion molecules. *Exp. Hematol.*, 1992, 20, 288-301.

[57] Verfaillie, C; Hurley, R; Bhatia, R; McCarthy, JB. Role of bone marrow matrix in normal and abnormal hematopoiesis. *Crit. Rev. Oncol. Hematol.*, 1994, 16, 201-224.

[58] Hynes, RO. Integrins: versatility, modulation, and signaling in cell adhesion. *Cell*, 1992, 69, 11-25.

[59] Ruoslahti, E. Integrins. *J. Clin. Invest.*, 1991, 87, 1-5.

[60] Kopan, R; Lee, J; Lin, MH; Syder, AJ; Kesterson, J; Crutchfield, N; Li, CR; Wu, W; Books, J; Gordon JI. Genetic mosaic analysis indicates that the bulb region of coat hair follicles contains a resident population of several active multipotent epithelial lineage progenitors. *Dev. Biol.*, 2002, 242, 44-57.

[61] Zhang, J,; Niu, C; Ye, L; Huang, H; He, X; Tong, WG; Ross, J; Haug, J; Johnson, T; Feng, JQ. et al. Identification of the haematopoietic stem cell niche and control of the niche size. *Nature*, 2003, 425, 836-41.

[62] Calvi, LM; Adams, GB; Weibrecht, KW; Weber, JM; Olson, DP; Knight, MC; Martin, RP; Schipani, E; Divieti, P; Bringhurst, FR. et al. Osteoblastic cells regulate the haematopoietic stem cell niche. *Nature*, 2003, 425, 841-6.

[63] Yokota, T; Oritani, K; Mitsui, H; Aoyama, K; Ishikawa, J; Sugahara, H; Matsumura, I; Tsai, S; Tomiyama, Y; Kanakura, Y. et al. Growth-supporting activities of fibronectin on hematopoietic stem/progenitor cells *in vitro* and *in vivo*: structural requirement for fibronectin activities of CS1 and cell-binding domains. *Blood,* 1998, 91, 3263-3272.

[64] Takakura, N; Huang, XL; Naruse, T; Hamaguchi, I; Dumont, DJ; Yancopoulos, GD; Suda T. Critical role of the TIE2 endothelial cell receptor in the development of definitive hematopoiesis. *Immunity,* 1998, 9, 677-686.

[65] Arai, F; Hirao, A; Ohmura, M; Sato, H; Matsuoka, S; Takubo, K; Ito, K; Koh, GY; Suda, T. Tie2/angiopoietin-1 signaling regulates hematopoietic stem cell quiescence in the bone marrow niche. *Cell,* 2004, 118, 149-61.

[66] Varnum-Finney, B; Purton, LE; Yu, M; Brashem-Stein, C; Flowers, D; Staats, S; Moore, KA; Le Roux, I; Mann, R. et al. The Notch ligand, Jagged-1, influences the development of primitive hematopoietic precursor cells. *Blood,* 1998, 91, 4084-4091.

[67] Varnum-Finney, B; Xu, L; Brashem-Stein, C; Nourigat, C; Flowers, D; Bakkour, S; Pear, WS; Bernstein, ID., Pluripotent, cytokine-dependent, hematopoietic stem cells are immortalized by constitutive Notch1 signaling. *Nat. Med.*, 2000, 6, 1278-1281.

[68] Karanu, FN; Murdoch, B; Gallacher, L; Wu, DM; Koremoto, M; Sakano, S; Bhatia M. The notch ligand jagged-1 represents a novel growth factor of human hematopoietic stem cells. *J. Exp. Med.*, 2000, 192, 1365-1372.

[69] Karanu, FN; Murdoch, B; Miyabayashi, T; Ohno, M; Koremoto, M; Gallacher, L; Wu, D; Itoh, A; Sakano, S; Bhatia, M. Human homologues of Delta-1 and Delta-4 function as mitogenic regulators of primitive human hematopoietic cells. *Blood,* 2001, 97, 1960-1967.

[70] Ohishi, K; Varnum-Finney, B; Bernstein, ID. Delta-1 enhances marrow and thymus repopulating ability of human CD34(+)CD38(-) cord blood cells. *J. Clin. Invest.*, 2002, 110, 1165-1174.

[71] Satoh, Y; Matsumura, I; Tanaka, H; Ezoe, S; Sugahara, H; Mizuki, M; Shibayama, H; Ishiko, E; Ishiko, J; Nakajima, K. et al. Roles for c-Myc in self-renewal of hematopoietic stem cells. *J. Biol. Chem.*, 2004, 279, 24986-24993.

[72] Reya, T; O'Riordan, M; Okamura, R; Devaney, E; Willert, K; Nusse, R; Grosschedl, R. Wnt signaling regulates B lymphocyte proliferation through a LEF-1 dependent mechanism. *Immunity*, 2000, 13, 15-24.

[73] Ivanova, NB; Dimos, JT; Schaniel, C; Hackney, JA; Moore, KA; Lemischka, IR. A stem cell molecular signature. *Science*, 2002, 298, 601-4.

[74] Willert, K;Brown, JD; Danenberg, E; Duncan, AW; Weissman, IL; Reya, T; Yates, JR, 3rd; Nusse R. Wnt proteins are lipid-modified and can act as stem cell growth factors. *Nature*. 2003, 423, 448-52.

[75]   Murdoch, B; Chadwick, K; Martin, M; Shojaei, F; Shah, KV; Gallacher, L; Moon, RT; Bhatia, M. Wnt-5A augments repopulating capacity and primitive hematopoietic development of human blood stem cells *in vivo. Proc. Natl. Acad. Sci USA*, 2003, 100, 3422-7.

[76]   Reya, T; Duncan, AW; Ailles, L; Domen, J; Scherer, DC; Willert, K; Hintz, L; Nusse, R; Weissman, IL. A role for Wnt signalling in self-renewal of haematopoietic stem cells. *Nature*, 2003, 423, 409-14.

[77]   Trowbridge, JJ; Xenocostas, A; Moon, RT; Bhatia, M. Glycogen synthase kinase-3 is an *in vivo* regulator of hematopoietic stem cell repopulation. *Nat. Med.*, 2006, 12, 89-98.

[78]   Kirstetter, P; Anderson, K; Porse, BT; Jacobsen, SE; Nerlov, C. Activation of the canonical Wnt pathway leads to loss of hematopoietic stem cell repopulation and multilineage differentiation block. *Nat. Immunol.*, 2006, 7, 1048-56.

[79]   Scheller, M; Huelsken, J; Rosenbauer, F; Taketo, MM; Birchmeier, W; Tenen, DG; Leutz, A. Hematopoietic stem cell and multilineage defects generated by constitutive beta-catenin activation. *Nat. Immunol.*, 2006, 7, 1037-47.

[80]   [81] Robbins, DJ; Nybakken, KE; Kobayashi, R; Sisson, JC; Bishop, JM; Therond, P. Hedgehog elicits signal transduction by means of a large complex containing the kinesin-related protein costal2. *Cell*, 1997, 90, 225-234.

[81]   Murone, M; Rosenthal, A; de Sauvage, FJ. Hedgehog signal transduction: from flies to vertebrates. *Exp. Cell Res.*, 1999. 253, 25-33.

[82]   Bhardwaj, G; Murdoch, B; Wu, D. Sonic hedgehog induces the proliferation of primitive human hematopoietic cells via BMP regulation. *Nat. Immunol.,* 2001, 2,178-180.

[83]   Wilson, A; Murphy, MJ; Oskarsson, T; Kaloulis, K; Bettess, MD; Oser, GM; Pasche, AC; Knabenhans, C; Macdonald, HR; Trumpp, A. c-Myc controls the balance between hematopoietic stem cell self-renewal and differentiation. *Genes Dev.,* 2004, 18, 2747-2763.

[84]   Melotti, P; Calabretta, B. The transcription factors c-myb and GATA-2 act independently in the regulation of normal hematopoiesis. *Proc. Natl. Acad. Sci. USA*, 1996, 93, 5313-5318.

[85]   Schmidt, M; Nazarov, V; Stevens, L; Watson, R; Wolff, L. Regulation of the resident chromosomal copy of c-myc by c-Myb is involved in myeloid leukemogenesis. *Mol. Cell. Biol.*, 2000, 20, 1970-1981.

[86]   Mucenski, ML; McLain, K; Kier, AB; Swerdlow, SH; Schreiner, CM; Miller, TA; Pietryga, DW; Scott, WJ, Jr; Potter, SS. A functional c-myb gene is required for normal murine fetal hepatic hematopoiesis. *Cell,* 1991, 65, 677-689.

[87]   Tsai, FY; Keller, G; Kuo, FC; Weiss, M; Chen, J; Rosenblatt, M; Alt, FW; Orkin SH. An early haematopoietic defect in mice lacking the transcription factor GATA-2. *Nature*, 1994, 371, 221-226.

[88]   Heyworth, C; Gale, K; Dexter, M; May, G; Enver, T. A GATA-2/estrogen receptor chimera functions as a ligand-dependent negative regulator of self-renewal. *Genes. Dev.* 1999, 13, 1847-1860.

[89]   Ezoe, S; Matsumura, I; Nakata, S; Gale, K; Ishihara, K; Minegishi, N; Machii, T; Kitamura, T; Yamamoto, M; Enver, T. et al. GATA-2/estrogen receptor chimera regulates cytokine-dependent growth of hematopoietic cells through accumulation of p21(WAF1) and p27(Kip1) proteins. *Blood*, 2002, 100, 3512-3520.

[90]   Kitajima, K; Masuhara, M; Era, T; Enver, T; Nakano, T. GATA-2 and GATA-2/ER display opposing activities in the development and differentiation of blood progenitors. *EMBO J.* 2002, 21, 3060-3069.

[91]   Persons, DA; Allay, JA; Allay, ER; Ashmun, RA; Orlic, D; Jane, SM. Enforced expression of the GATA-2 transcription factor blocks normal hematopoiesis. *Blood,* 1999, 93, 488-499.

[92]   Sauvageau, G; Thorsteinsdottir, U; Eaves, CJ; Lawrence, HJ; Largman, C; Lansdorp, PM; Humphries, RK. Overexpression of HOXB4 in hematopoietic cells causes the selective expansion of more primitive populations *in vitro* and *in vivo. Genes Dev.*, 1995, 9, 1753-1765.

[93]   Antonchuk, J; Sauvageau, G; Humphries, RK. HOXB4-induced expansion of adult hematopoietic stem cells *ex vivo. Cell*, 2002, 109, 39-45.

[94]   Buske, C; Feuring-Buske, M; Abramovich, C; Spiekermann, K; Eaves, CJ; Coulombel, L; Sauvageau, G; Hogge, DE; Humphries, RK. Deregulated expression of HOXB4 enhances the primitive growth activity of human hematopoietic cells. *Blood*, 2002, 100, 862-868.

[95] Knoepfler, PS; Sykes, DB; Pasillas, M; Kamps, MP. HoxB8 requires its Pbx-interaction motif to bl7ck differentiation of primary myeloid progenitors and of most cell line models of myeloid differentiation. *Oncogene,* 2001, 20, 5440-5448.

[96] Kroon, E; Thorsteinsdottir, U; Mayotte, N; Nakamura, T; Sauvageau, G. NUP98-HOXA9 expression in hemopoietic stem cells induces chronic and acute myeloid leukemias in mice. *EMBO J.,* 2001, 20, 350-361.

[97] Zhang, XB; Beard, BC; Trobridge, G; Hackman, R; Bryant, E; Humphries, RK; Kiem, HP. Development of Leukemia after HOXB4 Gene Transfer in the Canine Model. *ASH Annual Meeting,* 2006, [204]

[98] Krosl, J; Austin, P; Beslu, N; Kroon, E; Humphries, RK; Sauvageau, G. *In vitro* expansion of hematopoietic stem cells by recombinant TAT-HOXB4 protein. *Nat. Med.,* 2003, 9, 1428-32.

[99] Amsellem, S; Pflumio, F; Bardinet, D; Izac, B; Charneau, P; Romeo, PH; Dubart-Kupperschmitt, A; Fichelson, S. *Ex vivo* expansion of human hematopoietic stem cells by direct delivery of the HOXB4 homeoprotein. *Nat. Med.,* 2003, 9, 1423-1427.

[100] Tanaka, H; Matsumura, I; Itoh, K; Hatsuyama, A; Shikamura, M; Satoh, Y; Heike, T; Nakahata, T; Kanakura, Y. HOX decoy peptide enhances the *ex vivo* expansion of human umbilical cord blood CD34+ hematopoietic stem cells/hematopoietic progenitor cells. *Stem Cells,* 2006, 24, 2592-2602.

[101] Magli, MC; Largman, C; Lawrence HJ. Effects of HOX homeobox genes in blood cell differentiation. J *Cell Physiol* 1997, 173, 168-177.

[102] Buske, C; Humphries, RK. Homeobox genes in leukemogenesis. *Int. J. Hematol* 2000, 71, 301-308.

[103] Mahmoudi, T; Verrijzer, CP. Chromatin silencing and activation by Polycomb and trithorax group proteins. *Oncogene,* 2001, 20, 3055-3066.

[104] Park, IK; Qian, D; Kiel, M; Becker, MW; Pihalja, M; Weissman, IL; Morrison, SJ; Clarke, MF. Bmi-1 is required for maintenance of adult self-renewing haematopoietic stem cells. *Nature,* 2003, 423, 302-305.

[105] Jacobs, JJ; Kieboom, K; Marino, S; DePinho, RA; van Lohuizen, M. The oncogene and Polycomb-group gene bmi-1 regulates cell proliferation and senescence through the ink4a locus. *Nature,* 1999, 397, 164-168.

[106] Iwama, A; Oguro, H; Negishi, M; Kato, Y; Morita, Y; Tsukui, H; Ema, H; Kamijo, T; Katoh-Fukui, Y; Koseki, H. et al.Enhanced self-renewal of hematopoietic stem cells mediated by the polycomb gene product Bmi-1. *Immunity,* 2004, 21, 843-51.

[107] Rizo, A; Vellenga, E; de Haan, G; Schuringa, JJ. *Ex vivo* Expansion of Human Cord Blood CD34+ Cells by Overexpression of Bmi-1. *ASH Annual Meeting,* 2006, [1329]

[108] Roberts, JM. Evolving ideas about cyclins. *Cell,* 1999, 98, 129-132.

[109] Morgan, DO. Principles of CDK regulation *Nature,* 1995, 374, 131-134.

[110] Sherr, CJ; Roberts, JM. CDK inhibitors: positive and negative regulators of G1-phase progression. *Genes Dev.,* 1999, 13, 1501-1512.

[111] Brugarolas, J; Chandrasekaran, C; Gordon, JI; Beach, D; Jacks, T; Hannon, GJ; Radiation-induced cell cycle arrest compromised by p21 deficiency. *Nature,* 1995, 377, 552-557.

[112] Deng, C; Zhang, P; Harper, JW; Elledge, SJ; Leder, P. Mice lacking p21CIP1/WAF1 undergo normal development, but are defective in G1 checkpoint control. *Cell,* 1995, 82, 675-684.

[113] Nakanishi, M; Adami, GR; Robetorye, RS; Noda, A; Venable, SF; Dimitrov, D. Exit from G0 and entry into the cell cycle of cells expressing p21Sdi1 antisense RNA. *Proc. Natl. Acad. Sci. USA,* 1995, 92, 4352-4356.

[114] Taniguchi, T; Endo, H; Chikatsu, N; Uchimaru, K; Asano, S; Fujita, T; Nakahata, T; Motokura, T. Expression of p21(Cip1/Waf1/Sdi1) and p27(Kip1) cyclin-dependent kinase inhibitors during human hematopoiesis. *Blood,* 1999, 93, 4167-4178.

[115] Yaroslavskiy, B; Watkins, S; Donnenberg, AD; Patton, TJ; Steinman, RA. Subcellular and cell-cycle expression profiles of CDK-inhibitors in normal differentiating myeloid cells. *Blood,* 1999, 93, 2907-2917.

[116] Cheng, T; Rodrigues, N; Shen, H; Yang, Y; Dombkowski, D; Sykes, M; Scadden, DT. Hematopoietic stem cell quiescence maintained by p21cip1/waf1. *Science,* 2000 287, 1804-1808.

[117] Stier, S; Cheng, T; Forkert, R; Lutz, C; Dombkowski, DM; Zhang, JL; Scadden, DT. *Ex vivo* targeting of p21Cip1/Waf1 permits relative expansion of human hematopoietic stem cells. *Blood*, 2003, 102, 1260-1266.

[118] Nakayama, K; Ishida, N; Shirane, M; Inomata, A; Inoue, T; Shishido, N; Horii, I; Loh, DY; Nakayama, K. Mice lacking p27(Kip1) display increased body size, multiple organ hyperplasia, retinal dysplasia, and pituitary tumors. *Cell*, 1996, 85, 707-720.

[119] Kiyokawa, H; Kineman, RD; Manova-Todorova, KO; Soares, VC; Hoffmanm, ES; Ono, M; Khanam, D; Hayday, AC; Frohman, LA; Koff, A. Enhanced growth of mice lacking the cyclin-dependent kinase inhibitor function of p27(Kip1). *Cell*, 1996, 85, 721-732.

[120] Fero, ML; Rivkin, M; Tasch, M; Porter, P; Carow, CE; Firpo, E; Polyak, K; Tsai, LH; Broudy, V; Perlmutter, RM. et al. A syndrome of multiorgan hyperplasia with features of gigantism, tumorigenesis, and female sterility in p27(Kip1)-deficient mice. *Cell*, 1996, 85, 733-744.

[121] Cheng, T; Rodrigues, N; Dombkowski, D; Stier, S; Scadden, DT. Stem cell repopulation efficiency but not pool size is governed by p27(kip1). *Nat. Med.*, 2000, 6, 1235-1240.

[122] Yuan, Y; Shen, H; Franklin, DS; Scadden, DT; Cheng, T. *In vivo* self-renewing divisions of haematopoietic stem cells are increased in the absence of the early G1-phase inhibitor, p18INK4C. *Nat. Cell Biol.*, 2004, 6, 436-42.

[123] Furukawa, U; Kikuchi, J; Nakamura, M; Iwase, S; Yamada, H; Matsuda, M. Lineage-specific regulation of cell cycle control gene expression during haematopoietic cell differentiation. *Brit. J. Haemat.*, 2000, 110, 663-673.

[124] Nakayama, K; Nakayama, K. Cip/Kip cyclin-dependent kinase inhibitors: brakes of the cell cycle engine during development. *Bioessays*, 1998, 20, 1020-1029.

[125] Serrano, M; Lee, H-W; Chin, L; Cordon-Cardo, C; Beach, D; DePinho, RA. Role of the INK4a locus in tumor suppression and cell mortality. *Cell*, 1996, 85, 27-37.

[126] Ito, K; Hirao, A; Arai, F; Matsuoka, S; Takubo, K; Hamaguchi, I; Nomiyama, K; Hosokawa, K; Sakurada, K; Nakagata, N. et al. Regulation of oxidative stress by ATM is required for self-renewal of haematopoietic stem cells. *Nature*, 2004, 431, 997-1002.

[127] Janzen, V; Forkert, R; Fleming, HE; Saito, Y; Waring, MT; Dombkowski, DM; Cheng, T; DePinho, RA; Sharpless, NE; Scadden, DT. Stem-cell ageing modified by the cyclin-dependent kinase inhibitor p16INK4a. *Nature*, 2006, 443, 421-6.

[128] Teofili, L; Rutella, S; Chiusolo, P; La Barbera, EO; Rumi, C; Ranelletti, FO; Maggiano, N; Leone, G; Larocca, LM. Expression of p15INK4B in normal hematopoiesis. *Exp. Heamotol.*, 1998, 26, 1133-1139.

[129] Serrano, M; Lin, AW; McCurrach, ME; Beach, D; Lowe, SW. Oncogenic ras provokes premature cell senescence associated with accumulation of p53 and p16INK4a. *Cell*, 1997, 88, 593-602.

[130] Malumbres, M; Perez De Castro, I; Hernandez, MI; Jimenez, M; Corral, T; Pellicer A. Cellular response to oncogenic ras involves induction of the Cdk4 and Cdk6 inhibitor p15(INK4b). *Mol. Cell. Biol.*, 2000, 20, 2915-2925.

[131] Kamb, A; Gruis, NA; Weaver-Feldhaus, J; Liu, Q; Harshman, K; Tavtigian, SV; Stockert, E; Day, RS, 3rd; Johnson, BE; Skolnick, MH. A cell cycle regulator potentially involved in genesis of many tumor types. *Science*, 1994, 264, 436-440.

[132] [133] Ruas, M; Peters. G. The p16INK4a/CDKN2A tumor suppressor and its relatives. *Biochim. Biophys. Acta.*, 1998, 1378, F115-F177.

[133] Ogawa, S; Hangaishi, A; Miyawaki, S; Hirosawa, S; Miura, Y; Takeyama, K; Kamada, N; Ohtake, S; Uike, N; Shimazaki, C. Loss of the cyclin-dependent kinase 4-inhibitor (p16; MTS1) gene is frequent in and highly specific to lymphoid tumors in primary human hematopoietic malignancies. *Blood*, 1995, 86, 1548-1556.

[134] Haidar, MA; Cao, XB; Manshouri, T; Chan, LL; Glassman, A; Kantarjian, HM; Keating, MJ; Beran, MS; Albitar, M., p16INK4A and p15INK4B gene deletions in primary leukemias. *Blood*, 1995, 86, 311-315.

[135] Uchida, T; Kinoshita, T; Nagai, H; Nakahara, Y; Saito, H; Hotta, T; Murate, T. Hypermethylation of the p15INK4B gene in myelodysplastic syndromes. *Blood*, 1997, 90, 1403-1409.

[136] Shpall, EJ; Quinones, R; Giller, R; Zeng, C; Baron, AE; Jones, RB; Bearman, SI; Nieto, Y; Freed, B; Madinger, N. et al. Transplantation of *ex vivo* expanded cord blood. *Biol. Blood Marrow Transplant.*, 2002, 8, 368-376.

[137] Jaroscak, J; Goltry, K; Smith, A; Waters-Pick, B; Martin, PL; Driscoll, TA; Howrey, R; Chao, N; Douville, J; Burhop, S. et al. Augmentation of umbilical cord blood (UCB) transplantation with *ex vivo*-expanded UCB cells: results of a phase 1 trial using the AastromReplicell System. *Blood*, 2003, 101, 5061-7.

[138] Shpall E, de Lima M, Chan K, *et al.* Transplantation of Cord Bolld Expanded Ex Vivo with Copper Chelator. *Blood* 2004; 104: 281a.

In: Developments in Stem Cell Research
Editor: Prasad S. Koka

ISBN: 978-1-60456-341-2
© 2008 Nova Science Publishers, Inc.

Chapter 5

Autotransplantation of Purified Mensenchymal Stem Cells in a Patient with Acute Myocardial Infarction: A Case Study

Zhen-Xing Zhang[a,1],

Yan-Zhen Zhang[b,1], Li-Xue Guan[c], Yan-Ju Zhou[b], Xin-Xiang Zhao[b], Yao-Hong Cai[b], Li-Mei Zhao[b], Feng-Qi Li[b] and Long-Jun Dai [c,d,*]

[a]Department of Surgery, Weifang People's Hospital, Weifang Medical College, Weifang 261041, PR China

[b] Acute Care Unit, Weifang People's Hospital, Weifang Medical College, Weifang 261041, PR China

c Central Laboratory, Weifang People's Hospital, Weifang Medical College, Weifang 261041, PR China

[d] Department of Surgery, University of British Columbia, Rm 400 828 West 10th Ave., Vancouver, BC, Canada V5Z 1L8

ABSTRACT

Purified bone marrow-derived mesenchymal stem cells were autotransplanted to a patient with acute myocardial infarction. The employment of the sub-population of bone marrow-derived mononuclear cells was intended to clarify some disputable outcomes in heterogeneous mononuclear cell therapy. The improvement of myocardial perfusion and cardiac function was

---

[*]**Corresponding author**. Tel.: +1 604 875 4111x62501; fax: +1 604 875 4376. *E-mail address:* ljdai@interchange. ubc.ca (L.J. Dai).
[1]These authors contributed equally to this work

observed after delivery of mesenchymal stem cells through a combined procedure of primary intracoronary infusion and secondary intravenous infusion. This procedure is expected to enhance the engraftment efficacy of transplanted cells at infarcted myocardium.

Keywords: *acute myocardial infarction; mesenchymal stem cell; autotransplantation.*

1. INTRODUCTION

Since bone marrow-derived mesenchymal stem cells (MSCs) were discovered to have differentiative plasticity, great interest has been attracted on their potential therapeutic applications [1, 2]. There is increasing evidence indicating the therapeutic benefit of MSC transplantation in various disorders that are characterized by cell injury or cell loss, such as brain traumatic injury, stroke [3], Parkinson's disease [4], liver disease [5] as well as myocardial infarction [6,7]. In a recent multicenter clinical trial on acute myocardial infarction (AMI), Schachinger *et al* [8] reported an absolute improvement in the left ventricular ejection fraction (LVEF) with intracoronary infusion of BMC. By contrast, similar application of autologous BMC failed to show significant LVEF improvement in a smaller scale trial [9]. In another randomized, crossover clinical trial on chronic ischemic heart disease, Assmus *et al* [7] compared the efficacy of progenitor cells from different sources. Bone marrow-derived progenitor cell showed greater therapeutic benefit than the cells derived from circulating blood. Since total mononuclear cells isolated from density separation were used in all above mentioned clinical studies, the heterogeneous cell population was most likely the reason that apparently similar protocols yielded disparate outcomes. Rosenzweig [10] described this as "mixed results from mixed cells". In the present report, purified and characterized bone marrow-derived mesenchymal stem cells (MSCs) were autotransplanted into an AMI patient through a combined procedure, i.e., primary intracoronary infusion and secondary intravenous boost. Myocardial perfusion and cardiac function were assessed before and after cell therapy. The patient was regularly followed up for six months after cell transplantation. The therapeutic benefit was observed in both myocardial perfusion and cardiac function.

2. MATERIALS AND METHODS

2.1. Patient Enrolment

A 43 years old male patient suffered from chest pain and was short of breath for two weeks and was admitted into acute care unit. The patient was diagnosed with acute myocardial infarction and subjected to angioplasty and stent implantation followed by MSC administration. A written conformed consent was obtained from the patient. The protocol was approved by the Institutional Board of Medical Ethics at the hospital.

2.2. MSC Isolation and In Vitro Expansion

MSCs were isolated from mononuclear cell population in the bone marrow. The isolation method of mononuclear cells was previously described [11,12]. Briefly, under local anesthesia, bone marrow (10 ml) was aspirated from the posterior iliac crest using a heparinized syringe. Meanwhile, 30 ml of venous blood was taken for serum extraction. Bone marrow suspension was loaded on the top of an equal volume of Ficoll (density 1.077 g/ml, Sigma) preloaded centrifuge tube and centrifuged at 1800 rpm for 20 min. The top layer of mononuclear cells was collected and washed with Dulbcco modified Eagle's medium (DMEM, Gibco) three times. The isolated cells were suspended in DMEM/15% fetal bovine serum (FBS, Hang Zhou Biology Technology Company) supplemented with hPDGF (10ng/ml, Sigma) and hEGF (10ng/ml, Sigma), and then, seeded into 50 cm^2 flasks at a density of 10^6 cells/cm^2. The cells were cultured in a 95% humidified incubator at 37°C with 5% CO_2 in air. After 5 days, non-adherent cells were removed by replacing the medium. Attached cells developed into colonies within 3-7days. When these primary cultures of MSCs had reached 80-90% confluence, the cells were harvested using 0.25% trypsin (Gibco) and subcultured at a ratio of 1:3 for three passages. On the third passage, FBS and growth factors were removed from culture medium and replaced by 3% inactivated autologous serum.

2.3. Viability Assay, Microbiological Test and Cytogenetic Analysis

The cell viability was determined by trypan blue exclusion. Before transplantation, a sample of MSC suspension (10µl containing ~10^4 cells) was incubated with 5% Trypan blue at room temperature for 5 min. The percentage of non-stained cell number/total cell number was taken as the cell viability. Viability of ≥95% was one of the required criteria for the cell to be released for transplantation.

Gram staining and bacteria culture were taken as microbiologic tests. Negative Gram bacteria smear test and negative bacteria culture were required for MSC application.

After three passages of *in vitro* expansion, MSC cytogenetic analysis was performed by conventional staining and G-banding techniques as described in our previous report [12].

2.4. Stem Cell Marker Staining

To identify the stem cell population, harvested MSCs were stained with saturating amounts of monoclonal antibodies conjugated with fluorescein isothiocyanate (FITC): CD34-FITC (negative marker, Invitrogen) and CD44-FITC (positive marker, Invitrogen). The staining procedure was described in the manufacture's instruction. At least 20,000 events were analyzed by flow cytometry, E6155 (BD FACS Calibur).

2.5. Administration of MSCs in AMI patient

MSCs were administrated by direct intrcoronary infusion and intravenous infusion sequentially. Heparinization and filtration (40µm cell strainer, BD Falcon) were carried out to prevent cell clotting and microembolization during intracoronary and intravenous transplantation. For the primary administration, 6.08×10^7 MSCs in 20ml were given just after stent implantation via the same angioplasty balloon catheter. All cells were directly placed within the infarct-related artery. Two weeks later, the autologous transplantation was boosted by intravenous infusion of 3.42×10^8 MSCs.

2.6. Assessment of Myocardial Perfusion and Cardiac Function

The heart function was determined using 2 D Doppler Echocardiography (IE33, Philip). Myocardial perfusion was assessed using ^{99m}Tc-tetrofosmin single-photon emission computed tomography (SPECT, DS7, Sophy). All examinations were taken before and after MSC administration. The patient was also regularly followed up for six months.

3. RESULTS

3.1. Characterization of MSCs

In human adult bone marrow, stem cell population is mainly composed of haematopoietic stem cells (HSC) and a very small non-HSC portion consists of MSCs. CD34 is a surface marker for HSC and CD44 is one of the MSC markers. In the present study, over 95% *in vitro* expansed cells were CD34$^-$/CD44$^+$. MSC cytogenetic analysis showed normal diploid karyotype (46XY) and normal chromosome structure, which suggest that genetic stability was retained prior to autotransplantation.

3.2. The Changes of Heart Morphology and Cardiac Function

After the MSC transplantation, the internal diameters of all atria and ventricles became smaller compared with pre-transplantation (Table 1). Further changes were observed during follow up on different time points up to six months. Similar measurements were also taken for the diameters of aorta and pulmonary artery as well as the thicknesses of septum and posterior wall of left ventricle and no difference was detected between pre- and post-transplantation.

As shown in Table 2, all assessed cardiac function parameters were improved after MSC administration and showed the best performance at six months (6 m). Stroke volume (SV), fractional shortening (FS) and left ventricular ejection fraction (LVEF) were increased by 20%, 52% and 29% respectively after 6 months.

Table 1. Heart internal diameters (mm)

	Pre-Tx	Post-Tx		
		1 m	3 m	6 m
Right atrium	41	38	36	36
Right ventricle	30	28	27	26
Left atrium	46	37	36	35
Left ventricle	72	67	66	63

Tx: Transplantation

Table 2. Cardiac function assessments

	Pre-Tx	Post-Tx		
		1 m	3 m	6 m
SV (m l/m^2)	52	53.5	61	62
FS (%)	23	24	29	35
LVEF (%)	45	46	54	58

SV: Stroke volume; FS: Fractional shortening which was calculated as (LV end-diastolic diameter – LV end-systolic diameter)/LV end-diastolic diameter x 100%; LVEF: Left ventricular ejection fraction

3.3. The Improvement of Myocardial Perfusion

Figure 1 showed the myocardial perfusion status before cell therapy and the changes after MSC administration. The most apparent improvement was observed at one month (1m) especially in the YZ dimension.

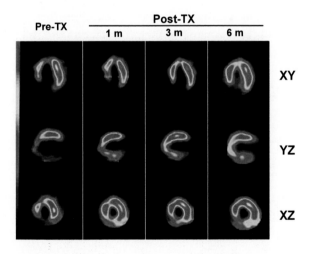

Figure 1. Myocardial perfusion at different time points before and after autologous MSC transplantation. The altitude of myocardial perfusion was detected by 99mTc-tetrofosmin single-photon emission computed tomography (SPECT) under rest conditions.

4. DISCUSSION

In the present report, a purified autologous MSC population was transplanted to an AMI patient. The cell preparation and administration protocol were different from recently reported clinical trials [7-9]. Bone marrow consists of a nonhomogeneous population of cells and contains a stem cell subpopulation. The stem cell fraction of bone marrow is mainly composed of hematopoietic stem cell (HSC). MSCs are non-HSC fraction and represent only 0.01-0.0001% of the nucleated cells in adult human bone marrow [13]. Through isolation and *in vitro* expansion, a purified and transplantable amount of MSCs were obtained in two weeks. The cells were delivered to the patient by a combined procedure, *i.e.* primary intracoronary infusion followed by secondary intravenous infusion. The safety and feasibility of intracoronary infusion have been verified by a number of clinical trials, and intravenous injection is a common method widely adopted in clinical and animal studies. Therefore, there was no safety concern on this combined procedure. The efficiency of cell therapy is determined by the homing number in the injured area. Freyman *et al* [14] quantitatively evaluated three methods of MSC delivery in a porcine myocardial infarction model. They rated MSC engraftment after MSC delivery through different methods as intracoronary infusion > endocardial injection > intravenous injection. Presumably, more MSCs could be homing to the infarcted area by this combined delivery procedure than any individual method alone. Further controlled study is required to confirm this point. However, this combined procedure represents a new way to deliver MSCs to the AMI patients.

The utilization of purified subpopulation of bone marrow nucleated cells could partially, if not completely, verify the disputable improvement of infarcted heart after bone marrow-derived cell therapy. In the present study, the most apparent improvement of myocardial perfusion was observed at one month after MSC transplantation, while the fast recovery of cardiac function was started two months later. This observation was consistent with the study conducted by Chen *et al* [15]. The early restoration of myocardial perfusion was most likely attributed to coronary angioplasty and stent placement, and the subsequent improvement in cardiac function was presumably owing to the repair of cardiomyocytes. Much slower functional recovery was expected without cell transplantation [15].The mechanism by which MSCs enhance the recovery of cardiac function was very complicated and far from clear. The transdifferentiation and, especially, paracrine-mediated myogenesis and angiogenesis at the infarcted area of the myocardium have been widely adopted in the field of MSC research [16, 17].

Recently, Müller-Ehmsen *et al* [18] conducted a study to compare the engraftment efficacy of bone marrow-derived mononuclear cells and MSCs in rat myocardial infarction model. They provided evidence that early engraftment was more efficient for MSCs than mononuclear cells, even though the difference disappeared in the mid-term, and long term persistence was poor for both cell types. As observed in our present study, the improved cardiac function was mainly due to MSC application. The combined delivery procedure would be capable of enhancing MSC engraftment efficacy of MSC in the infarcted area.

In conclusion, purified autologous mesenchymal stem cells were delivered to an AMI patient through a combined procedure of primary intracoronary infusion and secondary intravenous injection. This cell population was attributed to the recovery of myocardial perfusion and cardiac function. High MSC engraftment efficacy was expected through this combined cell deliver procedure.

ACKNOWLEDGMENTS

This work was supported by the Science and Technology Development Foundation of Weifang (2006-121). The authors would like to thank Dr. Jeff X. Zhou (N.I.H.) for his helpful suggestions and are grateful to Jing-Xian Wang, Rui-Li Wang, Cheng-Dong Wang, Hai-Bo Li and Xiao-Mei Ding for their technical assistance.

REFERENCES

[1]   Mezey E, Chandross KJ, Harta G, Maki RA, McKercher SR. Turning blood into brain: cells bearing neuronal antigens generated in vivo from bone marrow. *Science* 2000; 290:1779-82.

[2]   Jiang Y, Jahagirdar BN, Reinhardt RL, Schwartz RE, Keene CD, Ortiz-Gonzalez XR, *et al.* Pluripotency of mesenchymal stem cells derived from adult marrow. *Nature* 2002; 418:41-9.

[3]   Chopp M, Li Y. Treatment of neural injury with marrow stromal cells. *Lancet Neurol* 2002; 1:92-100.

[4]   Kim JH, Auerbach JM, Rodriguez-Gomez JA, Velasco I, Gavin D, Lumelsky N, *et al.* Dopamine neurons derived from embryonic stem cells function in an animal model of Parkinson's disease. *Nature* 2002; 418:50-56.

[5]   Kallis Y, Alison MR, Forbes SJ. Bone marrow stem cells and liver disease. *Gut* 2007; 56:716-24.

[6]   Strauer BE, Brehm M, Zeus T, Bartsch T, Schannwell C, Antke C, *et al.* Regeneration of human infarcted heart muscle by intracoronary autologous bone marrow cell transplantation in chronic coronary artery disease: the IACT Study. *J. Am. Coll. Cardiol.* 2005; 46:1651-58.

[7]   Assmus B, Honold J, Schachinger V, Britten MB, Fischer-Rasokat U, Lehmann R, *et al.* Transcoronary transplantation of progenitor cells after myocardial infarction. *N. Engl. J. Med.* 2006; 355:1222-32.

[8]   Schachinger V, Erbs S, Elsasser A, Haberbosch W, Hambrecht R, Holschermann H, *et al.* Intracoronary bone marrow-derived progenitor cells in acute myocardial infarction. *N. Engl. J. Med.* 2006; 355:1210-21.

[9]   Lunde K, Solheim S, Aakhus S, Arnesen H, Abdelnoor M, Egeland T, *et al.* Intracoronary injection of mononuclear bone marrow cells in acute myocardial infarction. *N. Engl. J. Med* 2006; 355:1199-209.

[10]   Rosenzweig A. Cardiac cell therapy--mixed results from mixed cells. *N. Engl. J. Med.* 2006; 355:1274-77.

[11]   Azizi SA, Stokes D, Auglli BJ, DiGirolamo C, Prockop DJ. Engraftmnt and migration of human bon marrow stromal cells implanted in the brains of albino rats – similarities to astrocyte grafts. *Proc. Natl. Acad Sci. USA* 1998; 95:3908-13.

[12]   Zhang ZX, Guan LX, Zhang K, Wang S, Cao PC, Wang YH, *et al.* Cytogenetic analysis of human bone marrow-derived mesenchymal stem cells passaged *in vitro*. *Cell Biol Int* 2007; 31:645-8.

[13]   Dazzi F, Ramasamy R, Glennie S, Jones SP, Roberts I. The role of mesenchymal stem cells in haemopoiesis. *Blood Rev.* 2006; 20:161-71.

[14]   Freyman T, Polin G, Osman H, Crary J, Lu M, Cheng L, *et al.* A quantitative, randomized study evaluating three methods of mesenchymal stem cell delivery following myocardial infarction. *Eur. Heart J.* 2006; 27:1114-22.

[15]   Chen SL, Fang WW, Ye F, Liu YH, Qian J, Shan SJ, et al. Effect on left ventricular function of intracoronary transplantation of autologous bone marrow mesenchymal stem cell in patients with acute myocardial infarction. *Am. J. Cardiol.* 2004; 94:92-5.

[16]   Haider HKh, Ashraf M._Bone marrow stem cell transplantation for cardiac repair. *Am. J. Physiol. Heart Circ. Physiol.* 2005; 288:H2557-67.

[17]   Giordano A, Galderisi U, Marino IR. From the laboratory bench to the patient's bedside: An update on clinical trials with mesenchymal stem cells. *J. Cell Physiol.* 2007; 211:27-35.

[18]   Müller-Ehmsen J, Krausgrill B, Burst V, Schenk K, Neisen UC, Fries JW, *et al.* Effective engraftment but poor mid-term persistence of mononuclear and mesenchymal bone marrow cells in acute and chronic rat myocardial infarction. *J. Mol. Cell Cardiol.* 2006; 41:876-84.

In: Developments in Stem Cell Research
Editor: Prasad S. Koka

ISBN: 978-1-60456-341-2
© 2008 Nova Science Publishers, Inc.

Chapter 6

Human Telomerase Catalytic Subunit (hTERT)-Transfected Mesenchymal Stem-Like Cells Derived from Fetal Liver Promote *Ex Vivo* Expansion of Hematopoietic Stem Cells

Yu-xia Yang[1, 2], Zhen-chuan Miao[1],
Hao-jian Zhang[1], Yun Wang[1]
and Mei-fu Feng[1*

[1]State Key Laboratory of Biomembrane and Membrane Biotechnology, Institute of
Zoology, Chinese Academy of Sciences, Beijing 100101, P.R. China
[2]Graduate University of Chinese Academy of Sciences, Beijing 100049, P.R. China

ABSTRACT

Fetal liver-derived adherent fibroblast-like stromal cells (FLSCs) were transfected with retrovirus, carrying the human telomerase catalytic subunit (*hTERT*) gene to establish an immortalized mesenchymal stem-like cell line, designated as FLMSL-hTERT. This cell line had the potential to differentiate into osteocytes and adipocytes. Their immunophenotype was similar with that of mesenchymal stem cells (MSCs) derived from bone marrow (BM), umbilical cord blood, skin, and adipose tissue. The cell line expressed mRNAs of SCF, Wnt5A, FL, KIAA1867, TGF-β, Delta-like, VEGF, SDF-1, PLGF and Jagged-1 that are known to promote hematopoiesis. The FLMSL-hTERT cells could dramatically support cord blood (CB) HSCs/HPCs expansion *ex vivo*, especially to maintain HSC properties *in vitro* at least for 8 wk; moreover, they appeared to be superior to primary human fetal liver stromal cells (FLSCs) in supporting *in vitro* hematopoiesis, because they were more potent than primary FLSCs in supporting the *ex vivo* growth of HSC/HPCs during long-term culture. The

---

* **Corresponding author:** Prof. Mei Fu Feng, State Key Laboratory of Biomembrane and Membrane Biotechnology, Institute of Zoology, Chinese Academy of Sciences, Beijing 100101, P.R. China. Tel: (8610) 64807562. Fax: (8610) 64807553. E-mail: fengmf@.ioz.ac.cn

FLMSL-hTERT cells have been maintained *in vitro* more than 150 population doublings (PDs) with the unchanged phenotypes. So this cell line may be of value for *ex vivo* expansion of HSCs/HPCs and for analyzing the human fetal liver hematopoietic microenvironments.

Keywords: *FLSCs; hTERT; FLMSL- hTERT; hematopoietic support.*

INTRODUCTION

Hematopoietic stem cells (HSCs) are rare, pluripotent cells that give rise to all hematopoietic cell types and possess extensive proliferative and self-renewal capabilities [1, 2]. The regulation of self-renewal and differentiation of HSCs requires a specific microenvironment of surrounding cells known as the stem cell niche [3, 4, 5, 6, 7], which contains hematopoietic-supporting stromal cells and extracellular matrix [8, 9]. During fetal development when hematopoiesis is transiently localized to the fetal liver (FL), there is a high frequency of cycling HSCs undergoing self-renewal, suggesting that the FL provides a conducive microenvironment for HSCs self-renewal [10, 11, 12, 13].

Stromal cells are thought to be essential in the regulation of hematopoieisis and maintaining the growth of HSC/HPCs [14]. However, relatively less proliferative potential of primary cells in addition to their low number has made the understanding of the niche very limited. Replicative senescence is a general phenomenon experienced by most somatic diploid cells due to the telomere loss during cell replication. Telomere activity is maintained by telomerase template, a catalytic subunit (human telomerase reverse transcriplate, hTERT), which has reverse transcriptase activity and associated with some protein components [15]. Ectopic overexpression of hTERT in somatic diploid cells could restore the telomerase activity and extended life span of the cells [16]. Indeed, some groups have already demonstrated that the ectopic overexpression of hTERT in primary cells restored the telomerase activity [17, 18, 19].

Previous studies have shown that mesenchymal stem cells (MSCs) could be isolated from human first-trimester fetal liver, readily expanded, and induced to differentiate in vitro [20]. We found that human FL contained the cells with an epithelial-to-mesenchymal transition (EMT) phenotype and these cells had the hematopoietic-surpportive ability [21]. Interestingly, in the present study, we obtained adherent, fibroblast-like stromal cells from human FL and transfected these cells with retrovirus containing hTERT gene successfully. We reported here the characteristics of this cell line and analyzed their ability to support *ex vivo* expansion of human CB CD34$^+$ HSCs/HPCs. The hTERT-transfected FL stromal cells, designated as FLMSL-hTERT, had the potential of mesenchymal stem-like cells, because they could differentiate into osteocytes and adipocytes under conditions that favored osteogenic and adipogenic differentiation of adult MSCs. The FLMSL-hTERT cells have been maintained *in vitro* more than 150 population doublings (PDs) with the stable phenotypes and without tumorigenicity. So this cell line may be useful for the study of mechanisms involved in stem cell-stromal cell interaction and the *ex vivo* expansion of human transplantable HSCs/HPCs.

MATERIALS AND METHODS

Materials

Human blood samples and fetal tissue collections for research purpose were approved by the Research Ethics Committees of Institute of Zoology, Chinese Academy of Sciences. National guidelines were complied with the regulation in relation to the use of fetal tissues only for research.

Antibodies against human antigens including CD29, CD34, CD44, CD45, CD31, CD58, HLA-DR, and the isotype antibodies were purchased from Becton Dickinson except for CD105, which is purchased from RandD systems. Cytokines about hematopoiesis including FL (10ng/mL), IL-6 (10ng/mL), SCF (10ng/mL) and TPO (10ng/mL) were from PEPRO TECH.

Isolation of Adherent and Fibroblast-Like Stromal Cells from Human Fetal Liver

Fetal liver samples were taken from 4 fetuses (mean age: 12 weeks). Fetal liver tissue was minced in Hank's balanced salt solution (HBSS) (Gibco BRL, Grand Island, NY, USA) and passed through wide bore syringe. Liver clumps were then digested by 0.10% (wt/v) collagenase IV (Gibco BRL, Grand Island, NY, USA) for 30 min at 37□ in a shaking bath. Then unselected nucleated cells from fetal liver samples were seeded on culture dishes coated with collagen and incubated overnight. After washing off the non-adherent cells, the adherent cells were cultured in D10F (Dulbecco's modified essential medium, DMEM) (Gibco BRL, Grand Island, NY, USA) supplemented with 10% fetal bovine serum (FBS) (Hyclone, USA), 10^{-4}M 2-mercaptoethanol (Fluka AG), 2mM L-glutamine, 5μg/ml insulin, 100U/ml penicillin and 100μg/ml streptomycin at 37°C under 5% CO_2 in a humidified atmosphere. After reaching confluence, the cells were harvested by 0.25% trypsin and 0.02% EDTA digestion and cryo-preserved as primary FLSCs, or passaged and used for gene transduction.

Retroviral Vector and Transduction of Primary Fetal Liver Stromal Cells

As described before [22], the pLPC-hTERT retroviral vector plasmid harboring a puromycin-resistant gene (Figure 1A) and the viral packaging plasmids containing gag-pol and env genes were kindly provided from Dr. Zhou Jian-Jun (Peking University, School of Life Science) and Dr. Gao Guang-Xia (Institute of Microbiology, Chinese Academy of Sciences), respectively. The 293T packaging cell line was co-transfected with pLPC-hTERT retroviral plasmid and the viral packaging plasmids in Lipofectanine 2000 (Invitrogen), following manufacturer's instruction. Viral supernatants were collected at 48 or 72 hr after transfection and then passed through a 0.45μm filter to remove cellular debris. FLSCs were plated into six-well plates with (the consistency of) $1 \sim 2.0 \times 10^5$ cells/well and cultured in dishes with D10F. The cells were washed twice with serum-free medium when they grew 70-80% confluent and transfected by adding 1.0 ml of retroviral supernatant in the presence of polybrene (8μg/ml), and incubated for 5 hr at 37 °C. Then, the transduction medium was replaced by 2.0 ml D10F, and puromycin (5μg/ml) was added into the culture 24 hr later. Puromycin-resistant cells grew out after 4 ~ 5 wk of culture, and then were digested and plated into new dishes in D10F for expansion.

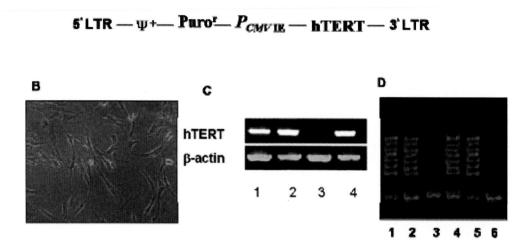

Figure 1. Establishment of hTERT-transduced FL-derived stromal cells. A, Schematic structure of the retroviral vectors containing hTERT. LTR indicates long terminal repeat; ψ, packaging signal; puro, puromycin-resistant gene; PCMV IE, cytomegalovirus (CMV) immediate early promoter; hTERT, human telomerase catalytic subunit. B, Morphology of hTERT-transduced FL-derived stromal cells under phase contrast microscope (×100). C, Constitutive expression of hTERT mRNA in the hTERT-transduced cells, but not FL primary stromal cells, demonstrated by RT-PCR. Lane 1, the hTERT-transduced cells at PDs of 75; Lane 2, the hTERT-transduced cells at PDs of 105; Lane 3, primary FL stromal cells at PDs of 10; Lane 4, Hela cells. D, Telomerase activity assessed by the stretch PCR method. Lane 1, the hTERT-transduced cells at PDs of 75; Lane 2, the hTERT-transduced cells at PDs of 105; Lane 3, primary FL stromal cells at PDs of 10; Lane 4, TSR8 as positive controls; Lane 5, Hela cells; and Lane 6, the CHAPS buffer alone. Data shown were from one of two reproducible experiments.

Adipogenic and Osteogenic Differentiation of the hTERT-Transfected Cells

Adipogenic differentiation was assessed by incubating the hTERT-transfected cell in IMDM supplemented with 1uM hydrocortisone, 0.5 mM 3-isobutyl-1-methylxanthine, 0.1 mM indomethacin and 10% rabbit serum (Sigma-Aldrich), and osteogenic differentiation was assessed by incubating the cells in IMDM supplemented with 0.1 uM dexamethasone, 0.2mM ascorbic acid and 10mM β-glycerol phosphate (Sigma-Aldrich) for 3 wk with medium changes twice weekly respectively as previously described [23]. Adipogenic differentiation was stained with fresh Oil-red-O solution (Sigma-Aldrich) and nontreated hTERT-transduced fibroblast-like cells were stained as controls. Alkaline phosphatase, an impoant marker for osteoblasts, was assayed with modified calcium-cobalt staining method [24] for evaluating osteogenic differentiation. Briefly, cultured cells were fixed with 95% alcohol for 10 min and stained for alkaline phosphatase activity as described [24]; the control slides were prepared by substituting H2O for β-glycerophosphate.

Table 1. Primer used in the experiment

Gene	Primer sequences	product (bp)	Size
β-actin	5'-TCA TGT TTG AGA CCT TCA A-3'	513bp	
	5'-GTC TTT GCG GAT GTC CAC G-3'		
LPL	5'-CCC TAC CCT TGT TAG TTA T-3'	453bp	
	5'-GTG TTT ATC AGA CCC TTT C-3'		
PPARγ2	5'-TGT CCA TAG AAG TCA CCC AT-3'	432bp	
	5'-TTT ATT TGC CAC AAC CCT-3'		
FABP4	5'-GCC AGG AAT TTG ACG AAG-3'	256bp	
	5'-ATC CCA CAG AAT GTT GTA GAG T-3'		
hOP	5'-GCT AAA CCC TGA CCC ATC CT-3'	424bp	
	5'-CAA CTC CTC GCT TTC CAT-3'		
hOC	5'-AAA GTT AAT GGG ATG GTC G-3'	415bp	
	5'-AAA GTT AAT GGG ATG GTC G-3'		
TGF-β	5'-CAA GTG GAC ATC AAC GGG TT -3'	297bp	
	5'-GCT CCA AAT GTA GGG GCA GG -3'		
SDF-1	5'-AAC GCC AAG GTC GTG GTC GTG CTG -3'	278bp	
	5'-CAC ATC TTG AAC CTC TTG TTT AAA AGC -3'		
KIAA1867	5'-CAG CCA AAA ATG ATA TCC GAC -3'	220bp	
	5'-GTG CTC TTT GAA GGT GAC -3'		
Delta-like-1	5'-CGG GAT CCC TCC ACA CAG ATT CTC CTG -3'	298bp	
	5'-CGG AAT TCT TAG ATC GGC TCT GTG CAG TAG -3'		
Shh	5'-ACT GGG TGT ACT ACG AGT CCA AGG -3'	211bp	
	5'-AAA GTG AGG AAG TCG CTG TAG AGC -3'		
Wnt5a	5'-ACA CCT CTT TCC AAA CAG GCC -3'	341bp	
	5'-GGA ATT GTT AAA CTC AAC TCT C -3'		
Jagged1	5'-GAT CCT GTC CAT GCA GAA CG -3'	440bp	
	5'-GGA TCT GAT ACT CAA AGT GG -3'		
PLGF	5'-CGA GTA CCC CAG CGA GGT G -3'	396bp	
	5'-GGA GTC ACT GAA GAG TGT GAC GG -3'		
KL/SCF	5'-GAC AGC TTG ACT GAT CTT CTG GAC -3'	360bp	
	5'-ACT GCT GTC ATT CCT AAG GGA GCT -3'		
TPO	5'- TGC TCC GAG GAA AGG TGC GTT-3'	460bp	

Table 1. (Continued)

Gene	Primer sequences	product Size (bp)
FL	5'- GGG AAG AGC GTA TAC TGT CCA-3' 5'-AAC AAC CTA TCT CCT CCT GCT -3'	307bp
VEGF	5'-GGC ACA TTT GGT GAC AAA GTG -3' 5'-TCG GGC CTC CGA AAC CAT GA -3'	649bp
hTERT	5'-CCT GGT GAG AGA TCT GGT TC -3' 5'-GGT GGA TGA TTT CTT GTT GGT G-3' 5'-AGG TGA GAC TGG CTC TGA TGG-3'	201bp

Reverse Transcriptase-Polymerase Chain Reaction (RT-PCR) Analysis

Total RNA was extracted from the cultured cells using the TRIzol Reagent (Life Technologies, Inc. Galthersburg, MD, USA), and reverse-transcribed into the first strand cDNA in a reaction primed by oligo (dT) primer. The RT reaction was performed for 1 hr at $42\,^{\circ}$C and stopped by heat inactivation for 5mins at $75\,^{\circ}$C. The cDNA samples were subjected to PCR amplification with specific primers under linear conditions. The primers used in the study are listed in Table 1. Following the amplification, each reaction mixture was visualized by 1.0% agarose gel electrophoresis. β-actin was served as an internal control.

Flow Cytometry

To analyze the surface markers of the cultured cells, cells grown in flasks were treated with trypsin and a total of $1{\sim}5{\times}10^5$cells/sample were collected for phenotype analysis of CD29, CD44, CD31, CD58, CD34, CD105, CD45 and HLA-DR. Each sample was first blocked with 0.1% BSA in PBS and then was stained with FITC-conjugated mAbs to CD29, CD44, CD105, CD45 or HLA-DR, or PE-conjugated mAbs to CD34, CD58 and CD31. FITC- or PE-conjugated mouse IgG1a (Becton Dickinson) was used as isotype controls. Cells were incubated at room temperature for 30mins and were analyzed by the flow cytometry (Becton Dickinson, San Jose, Ca, USA). 10,000 events were collected for each sample, and the data were analyzed with Cell-Quest software (Becton Dickinson, San Jose, CA, USA).

Analysis of the Telomerase Activity

Telomerase activity in the FLMSL-hTERT or primary FLSCs was determined using TRAPEZE Telomerase Detection Kit, according to manufacturer's instruction (Intergen Company) [25]. Briefly, the FLMSL-hTERT or primary FLSCs (about 10^6) were washed once in PBS and homogenized in 200μl ice-cold 1×CHAPS Lysis Buffer. The homogenate was incubated on ice for 30 min and then centrifuged at 12,000× g for 30 min at $4\,^{\circ}$C. The supernatant was collected and the protein content was determined by Coomassie brilliant blue assay. PCR reaction mixture (total 50μl) consisted of 39.6μl H_2O, 2μl (10~750ng/ul) cell extract, 5μl 10×TRAP reaction buffer, 1μl (50×) dNTP Mix, 1.0μl TS Primer (5'-AATCCGTCGAGCAGAGTT-3'), 1.0μl TRAP Primer Mix, and 0.4μl (5U/μl) Taq polymerase. The PCR reaction mixture was incubated at 30 °C in a thermal cycler for 30 min to extense TS primer by the telomerase. After elongation, the sample was heated at 94 °C for 3 min. This step inactivated the telomerase and released the CX primer after melting of the Ampli-wax gem. The PCR assay consisted of 30 cycles of 94 °C for 30 seconds, 50 °C for 30 seconds, and 72 °C for 45 seconds. TRAP reaction product (25μl) was analyzed by electrophoresis in 0.5×TBE buffer on 10% polyacrylamide non-denaturing gels. Gels were soaked in 0.5g/ml ethidium bromide in 1×TAE (Tris acetate EDTA) buffer for 30 min, viewed and photographed.

Isolation of Cord Blood CD34$^+$ Cells

Umbilical cord blood (CB) samples were obtained from normal full-term deliveries with informed consent (HaiDian Maternity Hospital, Beijing). CB samples containing heparin sodium salt (20U/ml) were doubly diluted in Phosphate Buffered Saline (PBS) with 0.6% Anticoagulants Citrate Dextrose-formula A (ACD-A) (6%ACD-A: 22.3g/L glucose, 22g/L sodium citrate and 8g/L citric acid in H_2O), and then centrifuged at 400g, 20 °C for 35 min over lymphocyte separation medium (1.077g/ml) (TBD Biotech, Tianjing, China). CB MNCs were collected from the interface of the separation medium and washed twice in PBS. CD34$^+$ cells were immuno-magnetically enriched from CB MNCs using the MACS CD34$^+$ Progenitor Cell Isolation Kit (Miltenyi Biotech Inc.), according to the manufacturer's instruction. The CD34$^+$ cell purity was 80~90%, determined by flow cytometry (FCM).

Expansion of Human CB CD34$^+$ Cells with the FLMSL-hTERT Cells in the Presence of Cytokines

Primary FLSCs or FLMSL-hTERT cells were irradiated at dose of 70 Gy (^{137} Cs source) and seeded in 24-well plates (5.0×10^5/well), and cultured overnight, washed twice with PBS, and cocultured with CD34$^+$ cells (2.0×10^4/well) in Iscove's Modified Dulbecco's Medium (IMDM) (Bio-WHITTAKER) supplemented with 10% FBS, 10^{-4}M 2-mercaptoethanol, 2mM L-glutamine, 5μg/ml insulin, 100U/ml penicillin and 100μg/ml streptomycin, and in the presence of a cytokine cocktail consisting of Flt ligand (FL) 10ng/ml, SCF 10ng/ml, thrombopoietin (TPO)10ng/ml and interleukin (IL)-6, 10ng/ml; all of them were purchased from Peprotech.

As a control, CD34$^+$ cells were cultured in the cytokine cocktail without presence of primary FLSCs or FLMSL-hTERT cells. After 10 d of culture, non-adherent and adherent hematopoietic cells that were weakly attached to stromal cells were harvested by gentle pipetting, counted, and analyzed for CD34 and CD38 expression by FCM.

Colony-Forming Unit (CFU) Assay

The colony-forming unit (CFU) assays were performed by plating the hematopoietic cells harvested from co-cultures into a complete methylcellulose medium (Stem Cell Technologies, Vancouver, BC, Canada) supplemented with the following growth factors: 50ng/ml SCF, 20ng/ml IL-3, 20ng/ml GM-CSF, 20ng/ml IL-6 and 3U/ml EPO (Kirin, Tokyo, Japan) at 37°C with 5%CO_2. After 14-16d culture, the number of colony-forming units (CFU-Cs) and lineage-specific colonies including colony-forming unit-erythrocyte (BFU-Es), colony-forming unit- granulocyte-macrophages (CFU-GMs) and colony-forming unit-mix (CFU-Mix's) were counted respectively.

Dexte-Modified Long-Term Culture Assay

For long-term culture analysis, 5×10^3 CB CD34$^+$ cells were plated into 6-well plates containing a nearly confluent, irradiated FLMSL -hTERT or primary FLSC monolayer in a-MEM supplemented with 12.5% horse serum (GIBCO BRL), 12.5% FCS (Hyclone) and 10^{-6} M hydrocortisone sodium hemisuccinate (Sigma) that was added just prior to use, using a modified procedure of Dexter et al. (26) 50% medium was replaced weekly with fresh medium. The cells were harvested at weeks of 4, 5, 6, 7 and 8 as described before and then subject to CFU assay. After 14 ~ 16 d of cultures, the number of CFU-Cs, BFU-Es, CFU-GMs and CFU-Mix's were counted to evaluate the frequency and multipotent differentiation of the long-term cultured cells.

Tumorigenicity Assay

To test tumorigenicity of FLMSL-hTERT cells, FLMSL-hTERT cells at PDs of 105 and Hela cells were injected s.c. into two-wk-old female Nu/Nu mice (n=3/group; 5.0×10^6/mouse) (obtained from Chinese Academy of Medical Science). Animals were inspected twice a week for tumor growth for fifteen wks.

Statistical Analysis

Results were expressed as mean ± SD. Statistical comparisons were performed using the two-tailed Student's t-test.

RESULTS

Establishment of the hTERT-Transduced Fibroblastlike Cell Line

We transfected FLSCs with retrovirus containing hTERT gene stably and obtained puromycin-resistent cells after 4-5 wk of puromycin selection. The resistent cells were cultured, morphologically and phenotypically characterized. The cells were adherent and fibroblastlike (Figure 1B). We first confirmed the expression of hTERT in the transfected stromal cells at population doublings (PDs) of 75 and 105 by RR-PCR. The results showed that hTERT mRNA could be detected in these cells, but not in primary FLSCs at PDs of 10 (Figure 1C). The level of hTERT mRNA in the resistent cells was comparable to that of HeLa cells (Figure 1C). Next, to evaluate if the hTERT mRNA is functional, we assessed the level of telomerase activity in the cells. As shown in Fig 1D, the telomerase activity was detected using a TRAPEZE Telomerase Detection Kit in the cells at PDs of 75 and 105. The activity does seem to be associated with the exogenous hTERT, because primary stromal cells did not exhibit telomerase activity. These results proved that exogenous hTERT gene that might have integrated into the genome of this cell line were functionally active. The doubling time (DT) of the hTERT-transfected fibroblast-like cell line was 4.5d similarly to that of primary FLSCs (5.5 d) within 12 wks of culture. While the DT of primary FLSCs became longer and finally died, the DT of the hTERT-transfected fibroblast-like cells remained unchanged in the morphology and phenotype. The line has been maintained for more than 150 population doublings (PDs) so far. Thus, we have successfully established an immortalized fibroblast-like cell line.

Differentiation of the hTERT-transfected Fibroblast-Like Cells

The differentiation potential of the hTERT-transduced fibroblast-like cells was tested by culturing cells under certain conditions that favored adipogenic and osteogenic differentiation of adult MSCs [23]. Treatment of the adipogenic medium resulted in adipocyte formation (Figure 2A), which was identified by the Oil red O staining (Figure 2B). When induced to differentiate under serum-free osteogenic conditions, the cells became flattened and broadened with increase in the time of induction (Figure 2D), and alkaline phosphatase activity was evident by the brown staining in the cytoplasm (Figure 2E). RT-PCR showed that adipocyte-specific genes (PPARγ2, peroxisome proliferative activated receptor, gamma 2, LPL, lipoprotein lipase and FABP4, fatty acid binding protein 4) and osteoblast-specific genes (hOSP, osteopontin and hOC, osteocalcin) were expressed in the corresponding induced cells, whereas they were not in the nontreated cells (Figure 2C, F). Thus, the hTERT-transfected fibroblastlike cells had the potential of mesenchymal stem cells, which were designateded as FLMSL-hTERT.

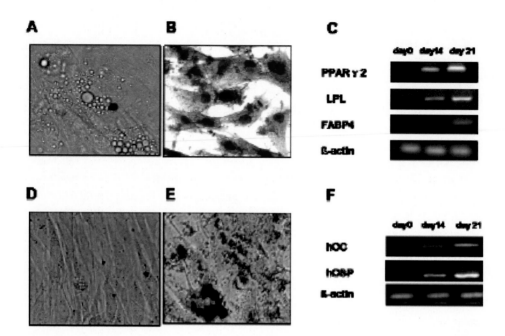

Figure 2. Differentiation potential of the hTERT-transduced FL-derived stromal cells. Adipogenic differentiation was evidenced by the formation of lipid vaculoes in phase-contrast photograph and the expression of specific genes. A. original magnification x 400; B. the lipid vacuoles stained by oil-red O; C. the expression of adipocyte-specific genes over 3 wk of induction. D. Osteogenic differentiation evidenced by morphology of cells after 3 wk of induction. E. Alkaline phosphatase expression, the brown staining was localized in the cytoplasm of hTERT-transduced cells demonstrated by Gomori modified calcium-cobalt staining, F. The expression of bone-specific genes over 3 wk of induction.

Immunophenotype of the FLMSL-hTERT Cells

The surface antigens on the hTERT-transfected stromal cells at PDs of 10, 75, and 105 were determined by flow cytometry. The results revealed that the cells were negative for CD34, CD45, indicating that these cells were nonhematopoietic origin (Figure 3). The cells were also negative for CD31 (PECAM-1) and HLA-DR (Figure 3). It was interesting that these cells were found to be positive for CD29 (β1-integrin), CD44 (the hyaluronate receptor), the intercellular adheresive molecular, CD58 and CD105 (endoglin). The immunophenotype of the FLMSL-hTERT cells was similar with that of mesenchymal stem cells (MSCs) derived from bone marrow, umbilical cord blood, skin and adipose tissue [27, 28, 29, 30]. Moreover, the results showed that the immunophenotype of the hTERT-transfected FL mesenchymal stem-like cells was stable even after cultured for long term in vitro.

Expression of Hematopoiesis-Related Factors by FLMSL-hTERT Cells

To test whether the FLMSL-hTERT cells could support hematopoiesis, we first examined the gene expression of hematopoiesis-related factors in these cells, including KIAA1867, Flt3 ligand (FL), stem cell factor/c-kit ligand (SCF/KL), thrombopoietin (TPO), vascular endothelial growth factors (VEGF), Placenta growth factor (PLGF), stromal cell-derived factor (SDF-1), Jagged-1, Delta-like1 (Dlk1), Sonic hedgehog (Shh), Wnt-5A, and transforming growth factor (TGF-β). As shown in Figure 4, the cell line constitutively expressed mRNAs of Wnt-5A, Dlk1 and Jagged-1, but not TPO and Shh. In addition, they also expressed mRNAs of KIAA1867, PLGF, VEGF, KL, FL, TGF-β and SDF-1.

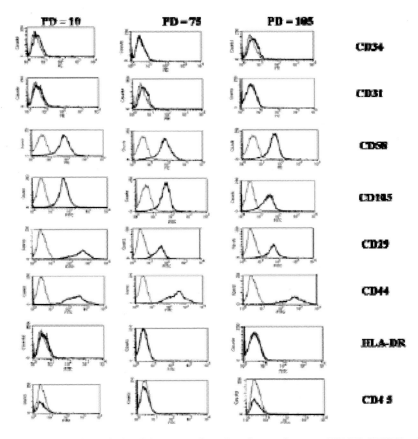

Figure 3. Flow cytometric analysis of the expression of surface antigens on FLMSL-hTERT cells at PDs of 10, 75, and 105. The FLMSL-hTERT cells at PDs of 10, 75, and 105 were stained with the FITC- or PE-conjugated monoclonal antibody to CD44 (H-CAM), CD29 (β1-integrin), CD58 LFA-3), CD45 (LCA), HLA-DR, CD31 (PECAM-1), CD105 (endoglin) or CD34, and analyzed by flow cytometry respectively. The antibodies were purchased from Becton Dickinson, except for CD105, which was from R&D systems. Dead cells were gated out by forward and side scatter. Data shown are from one representative experiment of three reproducible experiments.

Effect of the FLMSL-hTERT Cells on Ex Vivo Expansion of CB CD34+ Cells

We examined the effect of FLMSL-hTERT cells on the expansion of CD34+ enriched CB cells *in vitro*. 2.0×10^4 CD34+ enriched cells were cultured with or without a stromal layer (irradiated FLMSL-hTERT or primary FLSCs) in the presence of SCF, TPO, Flt3 and IL-6. After cultured for 10 d, the CB cells were harvested and tested for cell number, presence of primitive markers and hematopoietc progenitor capacity by CFU assay. As shown in Figure 5A, while CD34$^+$-enriched CB cells expanded by ~ 20 folds in the absence of stromal cells, they dramatically expanded by ~ 90 folds in the presence of FLMSL-hTERT or primary FLSCs. Importantly both CD34$^+$CD38$^-$ and CD34$^+$CD38$^+$cells were expanded dramatically in the presence of FLMSL-hTERT or primary FLSCs. The FLMSL-hTERT cells appear to be more potent than primary FLSCs (CD34$^+$ cells: 28.41 versus 24.81 folds; CD34$^+$CD38$^-$ cells: 23.37 versus 19.74 folds) in supporting HSC/HPC expansion. Consistently, although CD34$^+$ enriched CB cells cultured on FLMSL-hTERT have the similar progenitor capacity as those on primary FLSCs, CB cells cocultured on primary FLSCs or FLMSL-hTERT showed significantly higher progenitor capacity to CB cells cultured without stromal cells (Figure 5B). The results suggest that FLMSL-hTERT was as capable as freshly isolated primary FLSCs in expansion of HSCs/HPCs during short time.

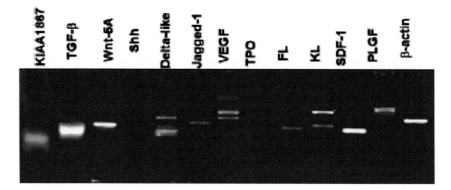

Figure 4. Cytokine profile expressed by the FLMSL-hTERT cells. Constitutive expression of hematopoiesis-associated cytokines in FLMSL-hTERT was analyzed by RT-PCR.

Effect of the FLMSL-hTERT Cells on Dexter-Type Long-Term Culture

To determine the capability of the FLMSL-hTERT cells in supporting self-renewal and maintaining multipotent differentiation of HSCs/HPCs, we performed long-term culture assay. CB CD34$^+$ cells were cocultured with FLMSL-hTERT cells or primary FLSCs for 4 ~ 8 wk without cytokines, and then were subject to CFU assay. After 14 ~ 16d of culture, the number of colonies with greater than 50 cells was counted under inverted microscope. As shown in Figure 6, both FLMSL-hTERT cells and the primary FLSCs could support the growth of the CFU-GM, CFU-Mix´s and BFU-E for 6 wk. While cocultured with FLMSL-hTERT cells, CB CD34$^+$ cells retained the capacity of multipotent differentiation lasting at least for 8 wk, because the number of CFU-GMs, CFU-E and CFU-Mix´s was not reduced at this time compared with at 4 wk of culture; in contrast, the CD34$^+$ cells cocultured with primary FLSCs progressively lost their multipotent differentiation with the time of culture, because CFCs were almost undetectable at 7 wk of coculture (Figure 6). Consistently, the number of total CFCs, CFU-Mix´s, BFU-Es and CFU-GMs from the cells cocultured with FLMSL-hTERT cells for 7 ~ 8 wk was significantly higher than that from the cells cocultured with primary FLSCs. The results suggest that the FLMSL-hTERT could efficiently maintain human HSC/HPCs self-renewal in the long-term culture. Moreover, FLMSL-hTERT cells are superior to primary FLSCs in supporting long-term culture of HSCs/HPCs.

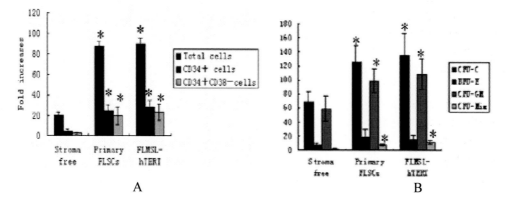

A   B

Figure 5. Effect of the FLMSL-hTERT cells on the expansion of CB CD34+ cells. 2.0×104 CB CD34+ were cocultured with or without a stromal layer (irradiated FLMSL-hTERT or primary FLSCs) in the presence of SCF, TPO, Flt3 and IL-6 for 10d, and analyzed by flow cytometry and CFU assay. A, Data indicate the fold increase compared with the initial number of cells. B, The number of the colonies, including BFU-Es, CFU-GMs and CFU-Mix´s , with greater than 50 cells was counted. Results are given as mean ± standard deviation (SD) of three separate experiments. *p<0.05 versus stromal cells free (n=3) (Student t test).

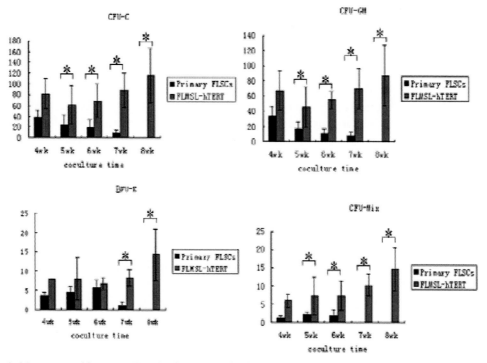

*P<0 .05, compared between FLMSL-hTERT and primary FLSCs (Student *t* test).

Figure 6. CFU count of cultured CB CD34+ in Dexter-modified long-term culture using the FLMSL-hTERT cells and primary FLSCs. 5.0×103 CD34+ CB cells were cocultured with hTERT-transduced or primary FLSCs without cytokines for 4 - 8 wk and then subject to CFU assay. After 14-16d of culture, the colonies, including BFU-Es, CFU-GMs, and CFU-Mix´s, with greater than 50 cells was counted. The results are expressed as mean ± SD (n=3).

Tumorigenicity Assay

To determine whether FLMSL-hTERT cells were malignant, we performed a tumorigenicity assay. The nude mice were injected s.c. with FLMSL-hTERT cells or HeLa cells. While all the mice injected with HeLa cells developed tumors, the mice injected with FLMSL-hTERT cells did not develop tumor within fifteen weeks of observation (n=3/group) (data not shown). The result indicates that FLMSL-hTERT cells were not tumorigenic.

DISCUSSION

Campagnoli et al [20] have reported that mesenchymal stem/progenitor cells could be obtained from human first-trimester fetal liver (FL), which are similar to those derived from adult bone marrow. However, the limited life span of the primary cells is the main obstacle for their further use. So we tried to transfect the primary FL stromal cells (FLSC) with hTERT gene to overcome cell crisis. However, whether immortalization of primary human cells could be achieved by the ectopic expression of hTERT might depend on cell types or culture conditions, and the ability of ectopic hTERT to extend life-span may also be related to the site of integration and the levels of telomere- or telomerase-associated proteins in a cell type-specific manner [31, 32, 33, 34]. In this paper, we reported the establishment and characteristics of mesenchymal stem-like cell line, through transfecting the primary FL adherent stromal cells with retrovirus containing hTERT. The cell line had the potential of mesenchymal stem cells (MSCs) because it could differentiate into osteocytes and adipocytes, which have been demonstrated by histochemical evaluations and RT-PCR analysis (Figure 2), under the conditions that favored adipogenic and osteogenic differentiation of adult MSCs [23]. Consistently, the FLMSL-hTERT cells were found to be positive for CD29 (β1-integrin), CD44 (the hyaluronate receptor), the intercellular adheresive molecular, CD58 and CD105 (endoglin) (Figure 3), which are similar with those reported for MSCs derived from other tissues [27, 28, 29, 30]. The expression of surface antigens on the FLMSL-hTERT cells kept unchanged (Figure 3) and the hTERT activity in these cells were stable (Figure 1) even after expanded for about two years in vitro.

Ectopic expression of hTERT in primary human cells may inhibit the replicative senescence, leading to immortalization. The immortalized cell lines such as FLMSL-hTERT do not necessarily undergo malignant transformation. Several reports have demonstrated that hTERT-transduced human cells develop into tumor at more than 100 PD [35, 36]. However, the tumorigenicity of the hTERT-transduced cell lines is inevitably conferred by the concomitant epigenetic alterations or oncogenic mutations [35, 36, 37]. Both the results of tumorigenicity assay and the fact that FLMSL-hTERT has been maintained in culture more than 150 PD without alteration of the phenotype, morphology and biological functions suggests that the cells have been immortalized, but not yet underwent significant oncogenic mutations.

FL is one of the important sites that HSCs emerge during ontogeny and can provide a conducive microenvironment to support HSCs [10, 11, 12, 13]. Here, we found that the FLMSL-hTERT cells expressed detectable levels of SCF, Wnt5A, FL, KIAA1867, TGF-β, Delta-like, VEGF, SDF-1, PLGF and Jagged-1, which are necessary in hematopoietic support, but not TPO and Shh. We further analyzed the function of the cell line in the HSC/HPC expansion. As expected, the FLMSL-hTERT can dramatically expand CB HSCs/HPCs ex vivo (Figure 5), especially maintaining HSC properties in vitro at least for 8 wk (Figure 6). The results also showed that the FLMSL-hTERT cells are superior to primary FLSCs in expanding HSCs/HPCs in vitro, especially for maintaining hematopoietic activity during long-term culture (Figure 6). The mechanisms underlying the enhanced expansion are not yet clear. However, this may be related to longer survival time of FLMSL-hTERT cells than primary FLSCs in the cultures.

In a conclusion, through transfection with retrovirus containing hTERT gene, we have developed an immortalized mesenchymal stem-like cell line derived from human fetal liver. The normal biological potential of these cells could be maintained. In this study, we confirmed and extended the findings that FL stromal cells form an important constituent of the hematopoietic niche and they are known to support self-renewal and proliferation of HSCs/HPCs. This cell line is superior to primary FLSCs in supporting HSC/HPC expansion in vitro. This newly established cell line might be of value for studying the mechanisms by which stromal cells regulate HSC/HPC expansion in FL and for developing a novel strategy for ex vivo expansion of human transplantable HSCs/HPCs.

ACKNOWLEDGEMENT

The work was supported by grants from National Natural Science Foundation of China (No. 30470427), the High–tech Project of the Chinese Ministry of Science and Technology (National "863" Project) (No. 2002AA205061), and the Chinese Academy of Sciences (No. KSCXZ-SW-322).

REFERENCES

[1] Morrison SJ, N Uchida and IL Weissman. (1996). The biology of hematopoietic stem cells. *Annu. Rev. Cell Dev. Biol.* 11:35-71.

[2] Ogawa M. (1993). Differentiation and proliferatioin of hematopoietic stem cells. *Blood* 81:2844-2853.

[3] Schofield R. (1978). The relationship between the spleen colony-forming cell and the haemopoietic stem cell. *Blood Cells* 4:7-25.

[4] Dexter TM, EG Wright, F Krizsa, and LG Lajtha. (1977). Regulation of haemopoietic stem cell proliferation in long term bone marrow cultures. *Biomedicine* 27:344-349.

[5] Zon LI. (1995). Developmental biology of hematopoiesis. *Blood* 86:2876-2891.

[6] Calvi LM, GB Adams, KW Weibrecht et al. (2003). Osteoblastic cells regulate the haematopoietic stem cell niche. *Nature* 425:841-846.

[7] Arai F, A Hirao, M Ohmura et al. (2004). Tie2/angiopoietin-1 signaling regulates hematopoietic stem cell quiescence in the bone marrow niche. *Cell* 118:149-161.

[8] Lanotte M, Allen TD, Dexter TM. Histochemical and ultrastructural characteristics of a cell line from human bone-marrow stroma. *J. Cell Sci.* 1981;50:281–297.

[9] Kobune M, Kawano Y, Ito Y, et al. Telomerized human multipotent mesenchymal cells can differentiate into hematopoietic and cobblestone area-supporting cells. *Exp. Hematol.* 2003;31:715–722.

[10] Murdoch B, L Gallacher, C Awaraji et al. (2001) Circulating hematopoietic stem cells as novel targets for in utero gene therapy. *FASEB J.* 10:1628-1630.

[11] Wilpshaar J, M Bhatia, HH Kanhai et al. (2002) Engraftment potential of human fetal hematopoietic cells in NOD/SCID mice is not restricted to mitotically quiescent cells. *Blood* 100:120-127.

[12] Ema H and H Nakauchi. (2000) Expansion of hematopoietic stem cells in the developing liver of a mouse embryo. *Blood* 95:2284-2288.

[13] Gallacher L, B Murdoch, DM Wu, FN Karanu, F Fellows and M. Bhatia. (2000) Identification of novel circulating human embryonic blood stem cells. *Blood* 96:1740-1747.

[14] Torok-Storb B, Iwata M, Graf L, Gianotti J, Horton H, Byrne MC. (1999) Dissecting the marrow microenvironment. *Ann. NY Acad. Sci.* 872:164-70

[15] Shay, J.W., Zou, Y., Hiyama, E. and Wright, W.E. (2001) Telomerase and cancer. *Hum. Mol. Genet* 10: 677-685.

[16] Simonsen, J.L., Rosada, C., Serakinei, N., Justesen, J., Stenderup, K., Rattan, S.I., Jensen, T.G. and Kassem, M. (2002) Telomerase expression extends the proliferative life-span and maintains the osteogenic potential of human bone marrow stromal cells. *Nat. Biotechnol.* 20: 592-596.

[17] Bodnar, A.G., Ouellette, M., Frolkis, M., Holt, S.E., Chiu, C.P., Morin, G.B., Harley, C.B., Shay, J.W., Lichtsteiner, S. and Wright, W.E. (1998) Extension of life-span by introduction of telomerase into normal human cells. *Science* 279: 349-352.

[18] Nisato, R.F., Harrison, J.A., Buser, R., Orei, L., Rinsch, C., Montesano, R., Dupraz, P. and Pepper, M.S. (2004) Generation and characterization of telomerase-transfected human lymphatic endothelial cells with an extended life span. *Am. J. Pathol.* 165: 11-24.

[19] Piera-Velazquez, S., Jimenez, S.A. and Stokes, D. (2002) Increased life span of human osteoarthritic chondrocytes by exogenous expression of telomerase. *Arthritis Rheum.* 46: 683-693.

[20] Cesare Campagnoli, Irene A. G. Roberts, Sailesh Kumar, Phillip R. Bennett, Ilaria Bellantuono, and Nicholas M. Fisk. (2001) Identification of mesenchymal stem/progenitor cells in human first-trimester fetal blood, liver, and bone marrow.

[21] Zhang H, Miao Z, He P, Yang Y, Wang Y and Feng M. (2005). The existence of epithelial-to-mesenchymal cells with the ability to support hematopoiesis in human fetal liver. *Cell Biol. Int.* 29: 213-219.

[22] Yuxia Yang, Zhenchuan Miao, Haojian Zhang, Yun Wang, Jianxin Gao and Meifu Feng. (2007). Establishment and characterization of a human telomerase catalytic subunit-transduced fetal bone marrow-derived osteoblastic cell line. *Differentiation.* 75:24–34.

[23] Lee OK, Kuo TK, Chen WM, Lee KD, Hsieh SL, and Chen TH. (2004) Isolation of multipotent mesenchymal stem cells from umbilical cord blood. *Blood.* 103:1669-1675.

[24] Quinones, J.A., van Bogaert, L.J. (1979) Nonspecific alkaline phosphatase activity in normal and diseased human breast. *Acta Histochem*, 64, 106-112.

[25] Lin, Y., Uemura, H., Fujinami, K., Hosaka, M., Harada, M. and Kubota,Y. (1997) Telomerase activity in primary prostate cancer. *J. Urol.*, 157: 1161-1165.

[26] Dexter TM, TD Allen and LG Lajtha. (1977). Conditions controlling the proliferation of hemapoietic stem cells in vitro. *J. Cell Physiol.* 91: 335-344.

[27] Pittenger MF, Mackay AM, Beck SC, et al. (1999) Multilineage potential of adult human mesenchymal stem cells. *Science.* 284:143–147.

[28] Erices A, Conget P, Minguell JJ. (2000) Mesenchymal progenitor cells in human umbilical cord blood. *Br. J. Haematol.* 109:235–242.

[29] Barry F, Boynton R, Murphy M, Haynesworth S, Zaia J. (2001) The SH-3 and SH-4 antibodies recognize distinct epitopes on CD73 from human mesenchymal stem cells. *Biochem. Biophys. Res. Commun.* 289:519–524.

[30] Young HE, Steele TA, Bray RA, et al. (2001) Human reserve pluripotent mesenchymal stem cells are present in the connective tissues of skeletal muscle and dermis derived from fetal, adult, and geriatric donors. *Anat Rec.* 264:51–62.

[31] Morales, C.P., Holt, S.E., Ouellette, M., Kaur, K.J., Yan, Y., Wilson, K.S., White, M.A., wright, W.E. and Shay, J.W. (1999) Absence of cancer-associated changes in human fibroblasts immortalized with telomerase. *Nat. Genet.* 21: 115–118.

[32] Bodnar, A.G., Ouellette, M., Frolkis, M., Holt, S.E., Chiu, C.P., Morin, G.B., Harley, C.B., Shay, J.W., Lichtsteiner, S., Wright, W.E. (1998) Extension of life-span by introduction of telomerase into normal human cells. *Science*. 279: 349-352.

[33] Kawano, Y., Kobune, M., Yamaguchi, M., Nakamura, K., Ito, Y., Sasaki, K., Takahashi, S., Nakamura, T., Chiba, H., Sato, T., Matsunaga, T., Azuma, H., Jkebuchi, K., Ikeda, H., Kato, J., Niitsu, Y. and Hamada, H. (2003) *Ex vivo* expansion of human umbilical cord hematopoietic progenitor cells using a coculture system with human telomerase catalytic subunit (*hTERT*)-transfected human stromal cells. *Blood*. 101: 532 540.

[34] Okamoto, T., Aoyama, T., Nakayama, T., Nakamata, T., Hosaka, T., Nishijo, K., Nakamura, T., Kiyono, T. and Toguchida, J. (2002) Clonal heterogeneity in differentiation potential of immortalized human mesenchymal stem cells. *Biochem. Biophys. Res. Commun.* 295: 354–361.

[35] Serakinci, N., Guldberg, P., Burns, J.S., Abdallah, B., Schrødder, H., Jensen, T. and Kassem, M. (2004) Adult human mesenchymal stem cell as a target for neoplastic transformation. *Oncogene* 23: 5095-5098.

[36] Lundberg, A.S., Randell, S.H., Stewart, S.A., Elenbaas, B., Hartwell, K.A., Brooks, M.W., Fleming, M.D., Olsen, J.C., Miller, S.W., Weinberg, R.A. and Hahn, W.C. (2002) Immortalization and transformation of primary human airway epithelial cells by gene transfer. *Oncogene* 21: 4577-4586.

[37] Stewart, S.A., Hahn, W.C., O'Connor, B.F., Banner, E.N., Lundberg, A.S., Modha, P., Mizuno, H., Brooks, M.W., Fleming. M., Zimonjic, D.B., Popescu, N.C., and Weinberg, R.A. (2002) Telomerase contributes to tumorigenesis by a telomere length-independent mechanism. *PNAS* 99: 12606-12611.

In: Developments in Stem Cell Research
Editor: Prasad S. Koka

ISBN: 978-1-60456-341-2
© 2008 Nova Science Publishers, Inc.

Chapter 7

Searching for Bone Marrow-Derived Cells in Experimental Models for Renal Injury: Reporter Systems Revisited

Martine Broekema, Martin C. Harmsen and Eliane R. Popa[*]

Department of Pathology and Laboratory Medicine, Medical Biology Section,
University Medical Center Groningen,
University of Groningen, the Netherlands

ABSTRACT

At this time point renal injury, ultimately resulting in renal failure, can not be therapeutically cured by endothelial, glomerular or tubular regeneration. Over the last five years, several reports appeared in which the replacement of damaged renal cells by bone marrow-derived cells (BMDC) was shown, thereby suggesting a therapeutic role for BMDC in renal regeneration. For correct interpretation of the function of these cells in renal repair, *in vivo* tracking of BMDC is crucial. Since various tracking methods with variable experimental outcomes have been reported, we will provide an overview of these methods and discuss their advantages and drawbacks for experimental renal disease models.

Keywords: *reporter, bone marrow-derived cells, detection, bone marrow transplantation, kidney.*

[*] **Corresponding author**: Eliane R. Popa, PhD. P.O. Box 196, 9700 AD Groningen, the Netherlands. Phone: +31 50 3615182, Fax: +31 50 3619911. E-mail: e.popa@med. umcg.nl

INTRODUCTION

At this time point renal injury, ultimately resulting in renal failure, can not be therapeutically cured by endothelial, glomerular and tubular regeneration. Bone marrow-derived cells (BMDC) might provide a therapeutic role in renal regeneration by replacement of damaged cells. In various models of renal injury, BMDC have been shown to be recruited from the circulation to the affected kidney, where they engraft and can adopt epithelial, endothelial or mesangial phenotypes, which suggests that damaged cells are actually replaced [1-6]. However, due to the low and variable number of BMDC that differentiated to renal-specific cell types in these models, the functional relevance of BMDC engraftment in renal repair is still subject of debate. One of the reasons for the variation in BMDC engraftment and differentiation may be the use of different BMDC tracking models. Moreover, there are reports suggesting that several BMDC tracking models are unreliable for identification of these engrafted cells *in vivo* [7;8]. Since BMDC tracking models are important for investigating possible therapeutic applications of BMDC in renal repair, we provide an overview and review the advantages and drawbacks of reporter systems used for tracing BMDC in the kidney (Table 1).

Reporter systems can generally be divided in two groups, i.e. those using genetic tracking devices that are present or newly introduced in the genome, and those using *ex vivo* labeling of cells.

Table 1. Reporter systems and their advantages and drawbacks

Reporter	Promoter	Advantages (+) and drawbacks (-) *	Circumvented by
Y-chrom.	-	+ Applicable in sex-mismatched patient material (13-16) - Unreliable in female to male combination - Underestimation due to thickness of sections (17) - Complicated, time-consuming (9) - Incorrect detection due to aggregates of fluorescent probe (9)	→ Correction factor
MHC/ Blood group	-	+ Applicable in mismatched patient material (6,14) + Simple detection techniques (14) - Mismatch can cause immune rejection (6)	
β-Gal	ROSA26	+ Enzymatic and immunohistochemical detection (9,30,31) + High sensitivity + Quantification using soluble color substrates + Ubiquitous renal expression (27) - Incorrect detection due to expression of mammalian- and senescence-associated β-Gal (9,28,29,32,33,35,36)	→ Adjust buffer pH → Use β-Gal antibody

Table 1. Reporter systems and their advantages and drawbacks (Continued)

Reporter	Promoter	Advantages (+) and drawbacks (-) *	Circumvented by
EGFP	chicken β-actin	+ Direct detection using microscopy or flow cytometry (38) + Detection in living cells (38) + Ubiquitous renal expression (38) + Rat reporter model available (5,46,47) + No endogenous EGFP expression (9) + Spectral variants of EGFP available (37,44,45) - Renal autofluorescence in the green spectrum (45) - EGFP in tissues often too weak for direct detection	→ Confocal microscopy → Use spectral variants → Use EGFP antibody
hPAP	ROSA26	+ Enzymatic and immunohistochemical detection (50,51) + Ubiquitous renal expression (26) + Stable expression in isolated cells in culture (52) + Rat reporter model available (26) - Endogenous alkaline phosphatase (48,49)	→ hPAP distinguished by heat stability
Firefly luciferase	hCMV immediate early gene	+ In vivo, real-time, non-invasive detection (53,55-57) + No animal-to-animal variation (56) + Ubiquitous renal expression (53) + Stable expression in isolated cells in culture (54) + High sensitivity - Exact localization can not be determined in vivo (57)	→ Ex vivo analysis on tissue samples
Epithelial-specific EGFP	Ksp-cadherin	+ Detection of tubular epithelial cells + Specific and permanent expression in mature tubular epithelium (42) + Additional phenotypical analysis not necessary - Mosaic expression pattern (42)	
Fibroblast-specific EGFP	FSP-1	+ Detection of fibroblasts + Specific and permanent expression in FSP-1+ cells (60) + Additional phenotypical analysis not necessary - Specificity FSP-1 debatable	

Table 1. Reporter systems and their advantages and drawbacks (Continued)

Reporter	Promoter	Advantages (+) and drawbacks (-) *	Circumvented by
Pro-COL1A2 specific luciferase/ β-Gal	pro-COL1A2	+ Detection of pro-collagen1A2 + Specific and permanent expression in pro-collagen 1-producing BMDC (35) + Additional analysis for Coll I production not necessary	
CFDA	-	+ Detection of ex vivo labeled cells + Label retained during development, meiosis, in vivo tracing, cell division, fusion (64) + Not transferred to adjacent cells (64) - Reduction of fluorescence level after cell division - Limited stability of the fluorescence, up to 20 days (66) - In vivo detection not possible (64) - Exact localization can not be determined in vivo	→ Expose kidney → Ex vivo analysis tissue samples (64)
Iron-dextran particles	-	+ Detection of ex vivo labeled cells + Non-invasive, in vivo, real-time detection with MRI (65) - MRI is laborious and requires equipment and trained personnel - Possible clearance via fagocytosis - Exact localization can not be determined in vivo	→ Ex vivo analysis on tissue samples (65)

* References are given between brackets.

BMDC TRACKING IN THE KIDNEY USING BONE MARROW TRANSPLANTATION

The most commonly used technique to track BMDC and study their role of BMDC after renal injury is transplantation of bone marrow from rodents that are genetically distinguishable from the recipient, into bone marrow-ablated recipients. After bone marrow transplantation, the transplanted bone marrow cells stably reconstitute the recipient's bone marrow compartment. In practice, the efficiency of bone marrow ablation and reconstitution with donor bone marrow is never 100% and a certain degree of underestimation of, for example the actual number of renal infiltrated BMDC, is always present. Another issue to consider is that bone marrow transplantation experiments with newly introduced genetically modified BMDC can elicit immunological rejection. However, the possibility of an interfering immune response has not been studied in bone marrow transplantation experiments described in this article.

Genetic "Tracking Devices"

Y Chromosome Detection

Male to female bone marrow transplantation offers a reliable, but laborious approach for tracking BMDC in the kidney. Several studies describe detection of male BMDC using Y-chromosome *in situ* hybridization (Y-ISH) to determine the bone marrow origin of cells engrafting the female kidney [2;4;9-12]. Y-chromosome detection is also used to study extra-renal, host-derived cells in tissue biopsy or autopsy material after sex-mismatched kidney transplantation [13-15], or to study renal infiltrated BMDC in patients after sex-mismatched bone marrow transplantation [16]. Thus, beside the use in animals after sex-mismatched bone marrow transplantation, Y-chromosome detection is used for recipient-derived cell detection in humans, in which the introduction of transgene-expressing cells, by for example bone marrow transplantation, is not possible.

The ISH method is laborious and may yield variable results, explaining the contradicting results of detection [2;4;9;10;12]. In a study by Duffield and colleagues [9], stating that post-ischemic tubular epithelial restoration occurs independently of BMDC, confocal laser fluorescent microscopy was used to show that Y^+ tubular cells often were artifacts. These artifacts were due to leukocytes overlying renal tubular structures, intratubular monocytes or nonspecific aggregates of fluorescent probe. This also suggests that the Y-chromosome positive tubular cells detected in previous studies in female mice following transplantation of bone marrow from male donor mice could be artifacts of imaging.

A drawback of this reporter system is that ISH is necessary to detect Y chromosome-positive cells in tissues. This technique time consuming and requires extensive pre-treatment of the sections, which affects tissue morphology and epitopes and therefore reduces the possibility to perform double-stainings for further characterization of the engrafted cells. Moreover, the method of Y-ISH for detection of BMDC can lead to an underestimation of Y-chromosome presence by distribution of the nucleus over multiple thin tissue sections, and loss of the Y chromosome for detection in some of the sections. Because of these bottlenecks, detection of the Y-chromosome in male tissue sections is never 100%, as it should be. Therefore, the exact number of Y chromosome-positive cells can only be estimated using a correction factor, as was done in a study by Direkze [17]. In this study radiation-induced injury was shown to elicit differentiation of BMDC to myofibroblasts in multiple organs, including the kidney. Using Y-ISH, the authors showed that only 50% of all male cells were Y chromosome-positive, which should have been 100%. Therefore, the observed number of renal Y chromosome-positive BMDC in the female kidney was corrected by dividing by 0.5 to estimate the exact number of renal BMDC. However, this approach is not very accurate since this is only an estimation and not the exact number of BMDC. Together with the fact that bone marrow transplantation is never 100% efficient, this estimation can lead to incorrect interpretation of data. Another limitation of sex-mismatched BMDC tracking, especially in the clinical setting, is that female to male detection is more difficult and unreliable, thereby excluding female to male transplantations for detection of recipient-derived cells. Therefore, the use of sex-mismatched bone marrow transplantation for tracking of BMDC can be regarded as a limited and insensitive method.

MHC and ABO Blood Group Antigen Detection

Although all reporter markers that are incorporated in the genome can be detected using *in situ* hybridization, other, less complicated methods for detection of recipient-derived cells in human tissue biopsies or post-mortem tissue are possible, e.g. MHC or ABO blood group antigen detection by immunohistochemistry [14]. Besides in patient studies, this strategy was also applied in rat studies, e.g. by Rookmaaker *et al.* [6], who used a rat allogenic BM transplant model in BN rats with WR rats as bone marrow donors to generate rat BM chimeras. In the model of anti-Thy-1.1-glomerulonephritis, BMDC were traced using a donor-specific major histocompatibility complex class-I monoclonal antibody and were found to differentiate towards glomerular endothelial and mesangial cells [6]. Although it was not reported in this study, possible immune rejection of MHC mismatched bone marrow and the necessity for immunosuppressive therapy can be a limiting factor of this detection method.

Transgenic Reporters

Transgenic Reporters under the Control of Ubiquitous Promoters

Tracing of BMDC in the kidney using reporter genes relies on the expression of the reporter gene (i.e. the transgene) by the BMDC. Expression of the reporter gene at all times can only be accomplished if the gene is driven by a promoter of a ubiquitously expressed gene, such as mouse metallothionein [18;19], β-actin from human [20], rat [21] and chicken [22], ubiquitin [23], simian virus 40 [24], cytomegalovirus immediate-early [25] or ROSA 26 [26].

β-Galactosidase

Bacterial β-galactosidase-transgenic mice expressing the *E.coli*-derived LacZ gene [27], are frequently used as a reporter for tracing BMDC. The presence of this reporter gene can be visualized by enzymatic detection of the bacterial β-galactosidase by X-gal staining or by immunohistochemical detection with an antibody against bacterial β-galactosidase. In the enzymatic detection, X-gal is converted by β-galactosidase to a blue reaction product which precipitates *in situ*. β-Galactosidase is among the most sensitive of reporter enzymes because only a few molecules of this enzyme readily convert X-gal in amounts that are detectable by light microscopy. An advantage of β-galactosidase is, that also soluble color substrates such as ONPG (o-nitrophenyl-β-D-galactopyranoside) exist that are employed to determine relative or total amount of reporter protein in tissue extracts.

Using this technique, previous studies have reported large numbers of X-gal[+] tubular cells after ischemic renal injury [1;2], thereby suggesting that BMDC play an important role in repair of renal injury. However, many mammalian tissues, including the kidney [28;29], contain endogenous β-galactosidase, an enzyme important for the enzymatic digestion of glycolipids [30;31]. This mammalian enzyme has an acidic pH optimum [32;33], whereas the bacterial β-galactosidase enzyme has a neutral pH optimum [34]. In most published protocols, a weak buffer such as phosphate-buffered saline (PBS) was used, which may have become acidic during the exposure of fixed tissues to X-gal. Moreover, after ischemia the kidney becomes acidic due to disturbed pH regulation. At low pH, endogenous β-galactosidase is detected and, thereby, false positive cells. Several studies indeed showed that, using the enzymatic detection method of β-galactosidase at low pH (6.5), endogenous β-galactosidase could not be distinguished from the bacterial β-galactosidase in injured kidneys [9][35]. Therefore, X-gal$^+$ tubular cells reported in previous studies [1;2] may have been detected as a result of increased intrinsic β-galactosidase activity, rather than the presence of β-galactosidase-positive bone marrow cells. Another disturbance in the detection of β-galactosidase is the presence of senescence-associated β-galactosidase (SA-β-gal), which is defined as β-galactosidase activity detectable at pH 6.0, in senescent cells [36].

Despite these detection problems, β-galactosidase can still be used as a reliable reporter, as long as bacterial β-galactosidase is clearly distinguished from endogenous mammalian β-galactosidase. This can be achieved by a simple modification of the X-gal method, raising the pH of the X-gal solution to weakly alkaline pH [29], or by using a commercially produced β-gal staining set which is designed to minimize staining from mammalian β-galactosidase [9]. Another method is based on immunolabeling with anti-bacterial β-galactosidase antibody instead of enzymatic detection of the reporter by X-gal staining. The study of Duffield [9] showed that, using anti-β-galactosidase, it was possible to discriminate between endogenous mammalian β-galactosidase activity and that resulting from LacZ gene expression. Therefore, β-galactosidase can be reliably detected without interference of endogenous mammalian β-galactosidase.

Enhanced Green Fluorescent Protein (EGFP)

Green fluorescent protein (GFP) is responsible for the green bioluminescence of the jellyfish *Aequorea Victoria*. Using this protein, transgenic mouse lines were generated, in which all tissues emitted green light under excitation. However, the first generation GFP transgenic mice proved to be unsuitable for use in experimental renal disease models, since GFP was, in the healthy animal, not expressed in all renal components [7; 37]. If, in the setting of GFP$^+$ bone marrow transplantation, BMDC regulate GFP in a similar way as kidney cells and BMDC differentiate to kidney cells, this would result in an underestimation of GFP expressing, bone marrow-derived, kidney cells. Therefore, mutants of the first generation of GFP transgenic mice were constructed that had about 35-fold brighter fluorescence, termed 'enhanced' GFP (EGFP) [37]. These mutants were the result of double amino acid substitutions in the wild-type GFP cDNA construct, placed under the control of the same chicken β-actin promoter and a cytomegalovirus enhancer as in the first generation of GFP transgenes. EGFP transgenic mice do express EGFP in all renal components, as well as in other tissues, with the exception of erythrocytes, hair [38] and some leukocytes (personal observation).

Chimeric mice reconstituted with EGFP bone marrow are commonly and successfully used to trace bone marrow-derived cells in renal disease models [39-43]. The advantage of EGFP as a reporter is that introduction of a substrate is not required, unlike other commonly used reporters such as β-galactosidase and alkaline phosphatase, allowing to monitor the presence of EGFP by sole illumination of tissue sections or living cells [38]. Moreover, EGFP is not disturbed by expression of endogenous EGFP and because the excitation optimum for EGFP is close to 488 nm, the transgenic cells can also be analyzed by flow cytometry. A drawback of EGFP is that in tissue sections EGFP is often too weak for direct fluorescence microscopy, making antibody labeling necessary for detection (personal observation, confirmed by other researchers). Moreover, the utility of EGFP as a reporter in renal disease models is limited by the fact that the kidney possesses intensive auto-fluorescence, which complicates detection of EGFP$^+$ cells, unless confocal microscopy is used. This problem can also be overcome by using one of the spectral variants of GFP, emitting blue, cyan or yellow light [37;44;45]. Moreover, these spectral variants of GFP can be very useful for achieving *in vivo* double-labeling.

Besides EGFP transgenic mice, rats expressing EGFP ubiquitously were generated. EGFP transgenic rats were originally established using the same construct and technique described for the production of EGFP transgenic mice [38]. Rats are, in comparison to mice, preferable in certain renal disease models, such as experimental kidney transplantation (easier microsurgical procedure) and anti-Thy1 antibody-mediated glomerulonephritis model (Thy1 is a well-established rat-specific mesangial marker, EGFP rat BM chimeras with Thy1 nephritis are described in references [5;46;47]). Therefore, the EGFP transgenic rat is an important tool for studying BMDC in renal disease.

Several characteristics of EGFP, its reliable detection, possibility of detection without introduction of a substrate in living cells and the availability of spectral variants of EGFP to avoid green autofluorescence or to simultaneously label multiple cell types, make this reporter an attractive option for use in BMDC tracking in renal disease models.

Human Placental Alkaline Phosphatase (hPAP)

Alkaline phosphatase (AP) dephosphorylates many types of phosphorylated molecules, for example nucleotides and proteins. Besides this functional property, which is extensively used in molecular biology, AP is used as a label in enzyme immunoassays. In humans, AP is present in all tissues throughout the body, but is particularly concentrated in liver, bile duct, kidney, bone and placenta. The advantage of human placental alkaline phosphatase (hPAP) is its heat stability, which allows to distinguish the placental form from other forms of AP [48;49].

Transgenic rats have been generated in which hPAP is placed under control of the ubiquitously active ROSA26 gene promoter [26]. The hPAP transgenic rat shows ubiquitous expression of hPAP in the kidney (Figure 1A) and is therefore suitable for BM transplantation experiments and subsequent tracking of BMDC in renal disease models.

We have used hPAP as a marker molecule to track BMDC in the post-ischemic rat kidney and showed BMDC differentiation towards tubular epithelial cells and myofibroblasts [50;51]. To allow for renal BMDC tracking, we reconstituted lethally irradiated F344 rats with ROSA26-hPAP transgenic bone marrow cells and subsequently studied infiltrating BMD (hPAP$^+$) cells in the kidney after unilateral ischemia/reperfusion injury. The heat-stability of the hPAP enzyme allowed reliable detection of hPAP$^+$ cells, without interference of endogenous alkaline phosphatase, which is abundantly present in the kidney. In our model, we showed that heat-inactivation of endogenous alkaline phosphatase resulted in complete absence of substrate conversion by this enzyme, without destroying the reactivity of hPAP (Figure 1B) [50;51].

Detection of hPAP$^+$ BMDC in renal tissue sections by (immuno)histochemical staining was confirmed by BMDC labeling and fluorescence-activated cell sorting (Figure 2). It has also been described that the expression of hPAP is stable in isolated cells in culture [52].

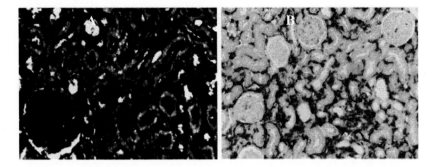

Figure 1. Renal expression of ROSA26-hPAP in transgenic rats and bone marrow chimeras. Expression of the ROSA26-hPAP gene was assessed by enzymatic hPAP staining on kidney sections. hPAP is ubiquitously expressed in the kidney of a healthy ROSA26-hPAP transgenic rat (A). ROSA26-hPAP expression can be easily detected on renal infiltrating hPAP+ BMDC (black) in ROSA26-hPAP bone marrow chimeric rats 7 days after induction of ischemia (B). No interference of endogenous AP was observed. Lens magnification 200x (B) and 400x (A).

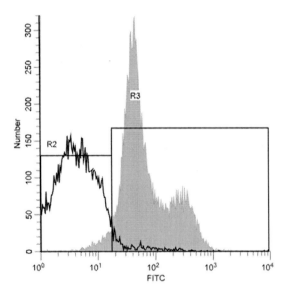

Figure 2. hPAP expression determined by FACS. ROSA26-hPAP transgenic bone marrow cells were detected by flow cytometry after isolation and labeling with anti-hPAP and a FITC labeled conjugate. The graph shows FITC expression in bone marrow cells of a non-transgenic F344 rat (left, transparent), and of a ROSA26-hPAP rat (right, red).

The stable and ubiquitous expression of hPAP, its applicability in rat models and simple detection methods, make the ROSA26-hPAP transgene a reliable reporter for studies on the fate of BMDC after renal injury.

Firefly Luciferase

Transgenic mice ubiquitously expressing luciferase from the North-American firefly (Photinus pyralis) have been generated, in which the firefly luciferase gene is controlled by promoter and enhancer elements of the human cytomegalovirus major immediate early gene [53]. Firefly luciferase, an enzyme that causes light emission from yellow to green wavelengths in the presence of a substrate (luciferin), oxygen, ATP and magnesium [54], is the most commonly used bioluminescent reporter in biomedical research. The fast rate of firefly luciferase enzyme turnover in the presence of the substrate luciferin and its short half-life allows for real-time measurements, because firefly luciferase does not accumulate intracellularly to the extent of other reporters [55]. Moreover, luciferase expression has been shown to be stable in isolated cells in culture [54].

The greatest advantage of the use of the bioluminescent firefly luciferase gene as a reporter protein is that the internal biological light source provided by luciferase can penetrate relatively easy through tissues, allowing *in vivo* detection using *in vivo* imaging techniques. Therefore, luciferase-produced bioluminescence can be non-invasively and repetitively measured in real-time in the same animal, thereby reducing the interference of animal-to-animal variation and requiring fewer animals per study [56].

A short-coming of the currently available techniques is that it is not possible to accomplish the imaging of real-time luciferase expression *within* individual cells *within* living organisms [57]. *In vivo* imaging can provide information on the renal localization of luciferase, but cellular localization must be determined, similar to other reporters, in (post-mortem) tissue sections or cell lysates. Nevertheless, sensitive noninvasive imaging of firefly luciferase gene expression makes this reporter suitable for studying BMDC recruitment to the circulation and homing to the injured kidney in a living organism.

Despite these advantages, bone marrow transplantation of ubiquitous luciferase-expressing bone marrow to study renal infiltrating BMDC has not been described. However, the use of a murine mesenchymal stem cell line transfected with a retroviral construct encoding firefly luciferase to study homing of these cells to the ischemic kidney was recently described (abstract by Kielstein J.T. *et al.* J. Am. Soc. Nephrol. 2007; 17: 527A). Detection of selectively luciferase-expressing BMDC was reported in the kidney [35] and will be discussed in the section on transgenic reporters under control of tissue or cell type-specific promoters. Moreover, the recently generated firefly luciferase/EGFP double transgenic mouse [58] will be very useful in studies on BMDC fate in the renal disease.

Transgenic Reporters under the Control of Tissue or Cell Type-Specific Promoters

Phenotypical changes or functional properties of renal infiltrating BMDC are mostly studied by combined immunohistochemistry detecting the reporter marker in conjunction with a cell type-specific marker. Another option is to perform bone marrow transplantation with transgenic bone marrow in which the reporter gene is driven by a cell type-specific promoter. Since differentiation of BMDC to that specific cell type will elicit activation of the promoter and expression of the reporter marker, this allows, in addition to mere localization of the BMDC, to evaluate potential differentiation of the BMDC to a certain cell type in a more reliable way. A problem that can be encountered when using these tissue or cell type-specific promoters is that activation of the promoter and thus expression of the reporter is not as specifically as it should be. However, when expression is cell-type-specific within the organ of interest, the reporter can be utilized. For example, in the first version of the GFP expressing 'green' mice, GFP was expressed in the renal podocytes, skeletal muscle, pancreas and heart [37]. Although GFP expression was not exclusively observed in renal podocytes, these mice could be used to study differentiation of BMDC to podocytes [7;59].

Epithelial-Specific Expression of EGFP

Lin *et al.* [42] crossed two mouse strains to accomplish conditional tubular epithelial specific expression of EGFP to study the contribution of intra-renal and BMDC to post-ischemic tubular regeneration. The first mouse strain, *Z/EG*, is a double reporter mouse. The first reporter, *lacZ,* is linked to a ubiquitous promoter and is flanked by two *loxP* sites. The second reporter, EGFP, resides further downstream. The EGFP gene will only be activated in the presence of cre recombinase, resulting in recombination between the *loxP* sites and subsequent deletion of the *lacZ* sequence. Once activated, EGFP expression is irreversible and inheritable, irrespective of the continued presence of cre recombinase.

The second transgenic mouse strain, cre^{ksp}, expresses cre recombinase under the control of the renal tubular epithelial-specific Ksp-cadherin promoter. Crosses between these two mouse strains result in the cre^{ksp}*;Z/EG* transgenic mice which specifically and permanently express EGFP in mature tubular epithelial cells [42]. In cre^{ksp}*;Z/EG* bone marrow chimeric mice, differentiation of BMDC to tubular epithelial cells would be visible by EGFP expression. However, the cre^{ksp}*;Z/EG* transgenic mice that were used in this study showed a mosaic expression pattern of EGFP, due to inefficient cre/loxP recombination and non-ubiquitous expression of the Z/EG promoter. Therefore BMDC fate could not be determined reliably using this reporter marker [42].

Fibroblast-Specific Expression of EGFP

To determine the source of renal interstitial fibroblasts in renal fibrosis, Iwano *et al.* [60] used FSP1.GFP bone marrow chimeras. In these chimeras, renal infiltrating BMDC express GFP under control of the FSP1 (fibroblast specific protein 1) promotor, and therefore, express GFP upon conversion to a fibroblast phenotype. Although Iwano demonstrated differentiation of BMDC to FSP-1$^+$ cells after unilateral ureter obstruction (UUO), the question remains if FSP-1 is specifically expressed by fibroblasts.

The use of transgenic mice constitutively expressing a reporter molecule under the control of an endothelial-specific promoter for bone marrow transplantation studies has, to the best of our knowledge, not been described in renal disease models. In cardiovascular research however, the use of bone marrow chimeras in which β-galactosidase expression is transcriptionally regulated by endothelial-specific promotors Flk-1 or Tie-2 has been proven suitable to study BMDC differentiation to an endothelial phenotype [61]. Moreover, the use of transgenic mice in which a reporter gene is placed under the control of a podocyte-specific promoter, such as nephrin or podocin, will also be a useful tool to extend our knowledge on BMDC differentiation in renal injury models.

Pro-Collagen 1A2-Specific Expression of Luciferase/B-galactosidase

The use of reporters driven by cell-type specific promoters for BMDC detection can give information about differentiation fates of BMDC. However, another important question is whether BMDC are functional in their differentiated state. This question was addressed in a study by Roufosse [35], using the UUO model of renal fibrosis to study the functional contribution of BMDC to fibrosis by determination of their capacity to produce collagen. To this end, a transgenic mouse was generated that expressed both luciferase and β-galactosidase reporter genes under the control of a promoter and enhancer element of the gene encoding pro-COL1A2 (coding for the α2 chain of the pro-collagen type 1). Roufosse demonstrated the unreliability of β-galactosidase in this model (see also β-galactosidase section). However, detection of pro-collagen 1 could still be accomplished by luciferase. The presence of luciferase was determined by measurement of luminescence or luciferase protein, or by *in situ* hybridization for luciferase mRNA, the latter allowing the authors to determine the exact location of luciferase activity and co-expression with other markers [35].

In the study of Roufosse, *in vivo* imaging techniques were not used to determine luciferase activity. When performed on living animals *in vivo* bioluminescence imaging would have provided information on renal luciferase activity and would have allowed for non-invasive tracing of luciferase activity in time in the same animal.

Stability of Transgene Expression

It may occur that a transgene is not expressed, that expression disappears in subsequent generations of the transgenic rodent, despite the presence of the transgene in the genome, or that expression is or becomes variable in different tissues. This phenomenon can be caused by gene silencing, which causes the loss of transcription of the particular gene [62]. Gene silencing can be the result of DNA methylation and/or histon deacetylation, causing alterations in chromatin structure by as yet unknown but cell-type restricted mechanisms. Both cause the shutting-off of gene transcription [63].

Gene silencing can also take place on a post-transcriptional level [62]. This occurs when mRNA of the transgene is degraded or blocked prior to translation e.g. by microRNA. Alternatively, the stability of transgenic protein may be affected, causing a high degradation rate, e.g. by increased ubiquitination.

The possibility of gene silencing in transgenic bone marrow transplantation models is often not considered and is likely to be an underestimated problem.

EX VIVO LABELING OF BMDC

Besides the use of bone marrow transplantation to study the role of BMDC in renal disease models, local or systemic infusion of BMDC, or a sub-set of BMDC, has been used. This approach does not require total body irradiation and is more relevant in pre-clinical, rather than experimental, studies. However, since the native bone marrow is not replaced by bone marrow transplantation, the infused BMDC will compete with physiologically recruited BMDC, which might hamper the interpretation of BMDC function.

Infusion of BMDC can be performed using the same genetically labeled reporter cells as used in bone marrow transplantation models, i.e. Y-chromosome, β-galactosidase, EGFP and hPAP. To allow detection of the infused cells in the kidney, non-genetic labeling methods have been applied, e.g. infusion of CFDA fluorescence labeled [64] or iron-dextran labeled cells [65].

Long-Lasting Dyes - CFDA

In a study by Tögel [64], carboxy-fluorescein diacetate (CFDA; Vybrant cell tracer kit, Molecular Probes) pre-labeled mesenchymal stem cells were infused in the left carotid artery to study their contribution to post-ischemic renal recovery. CFDA enters the cell by passive diffusion and starts to fluoresce after conversion by intracellular esterases, and is therefore suitable for marking living cells. The label is retained during development, meiosis and *in vivo* tracing, and is inherited by daughter cells after division or fusion, but is not transferred to adjacent cells in a population. However, after each cell division the fluorescence level is two-fold reduced. The major drawback of this labeling method and other comparable fluorescent labels is the limited stability of the fluorescence. *In vivo* the label could be traced up to 20 days after labeling [66] and is, therefore, not suitable for use in longer-term experiments. The emitted fluorescence of CFDA does not penetrate deep enough through tissue to allow for *in vivo* detection of the label. However, when the kidney was exposed and renal vasculature was visualized using rhodamine-dextran or FITC-albumin, BMDC within the renal vasculature could be observed *in vivo* using two-photon confocal microscopy [64].

Particles - Iron-Dextran Labeled Particles

Lange *et al.* [65] infused mesenchymal stem cells, pre-labeled with carboxy-dextran-coated iron oxide nanoparticles ("Resovist", Schering), into the thoracic aorta via a carotid artery after ischemia. These non-toxic iron-dextran particles are smaller than erythrocytes and are readily taken up by rat mesenchymal stem cells [65]. An advantage of this method is the non-invasive detection of the labeled cells by Magnetic Resonance Imaging (MRI), which allows for real-time imaging in the living animal. However, MRI is a laborious technique which requires equipment and trained personnel. Besides *in vivo* detection with MRI, *ex vivo* histological identification of the iron-labeled cells by Prussian blue staining can be used to determine the exact location of the cells. It is unknown how these iron-dextran particles, likely to be cleared by phagocytosis, influence the inflammatory response. In the study by Lange [65], the particles did not elicit an immune response, were stable and detectable in a non-invasive way, therefore, providing an elegant method for *ex vivo* labeling of cells.

SUMMARY AND FUTURE PERSPECTIVES

Various detection methods have been described to study the functional relevance of renal infiltrating BMDC. Provided that the essential prerequisites of enzymatic staining are met, all above mentioned reporters would seem applicable and useful for tracing BMDC in renal disease models. Some have, however, advantages over others. EGFP and luciferase both can be detected in living cells, which is an advantage for functional studies. The advantage of EGFP as a reporter molecule over luciferase is that detection of the protein can be accomplished without the introduction of a substrate. Luciferase needs a substrate for excitation, but it does not require external light excitation and, therefore, is less susceptible to background noise than EGFP. The background noise of EGFP, which is especially problematic in the kidney, can be overcome by the use of one of its spectral variants (cyan-, blue- and yellow fluorescent protein). These spectral variants also allow simultaneous labeling of multiple cell types. Both luciferase and EGFP have their advantages and disadvantages. However, a considerable disadvantage of EGFP is that, in contrast to firefly luciferase, it does not penetrate tissues enough for non-invasive *in vivo* detection of the signal in the living animal. Firefly luciferase does provide the opportunity to *in vivo*, real-time, repetitive measurements in the same animal. Where reporter molecules previously limited investigators to study the presence of BMDC only *ex vivo*, bioluminescent imaging (used for luciferase detection), magnetic resonance imaging (MRI) and Positron Emission Tomography (PET) techniques now allow for *in vivo* tracking of BMDC in living animals. Advantage of bioluminescent imaging over MRI and PET is that the use of light for bioluminescent imaging is safe and not subject to radioactive decay, as are the tracers used in PET, and does not require the introduction of particles or contrast fluids, as are required for MRI [56]. A major short-coming of the currently available imaging techniques is that their resolution is not sufficient for visualization of individual cells or nephrons [57]. Therefore, additional phenotypical analysis on tissue section will be necessary to study differentiation fates of BMDC.

In conclusion, most described reporter molecules can, knowing their strong and weak sides, be reliably applied for detection of renal infiltrating BMDC. The application determines which reporter is most suitable. Firefly luciferase would be suitable for *in vivo* detection of BMDC mobilization and homing to the damaged kidney. However, for phenotypical analysis of BMDC to determine their functional relevance in renal repair, EGFP has been described as the reporter marker with the most opportunities.

ACKNOWLEDGEMENTS

We would like to thank Dr. E. Sandgren for providing ROSA26-hPAP rats. This work was supported by grant C02.2031 and C05.2159 from the Dutch Kidney Foundation.

REFERENCES

[1]   Kale S, Karihaloo A, Clark PR *et al*: Bone marrow stem cells contribute to repair of the ischemically injured renal tubule. *J. Clin. Invest* 112:42-49, 2003.

[2]   Lin F, Cordes K, Li L *et al*: Hematopoietic stem cells contribute to the regeneration of renal tubules after renal ischemia-reperfusion injury in mice. *J. Am. Soc. Nephrol.* 14:1188-1199, 2003.

[3]   Morigi M, Imberti B, Zoja C *et al*: Mesenchymal stem cells are renotropic, helping to repair the kidney and improve function in acute renal failure. *J. Am. Soc. Nephrol.* 15:1794-1804, 2004.

[4]   Poulsom R, Forbes SJ, Hodivala-Dilke K *et al*: Bone marrow contributes to renal parenchymal turnover and regeneration. *J. Pathol.* 195:229-235, 2001.

[5]   Ito T, Suzuki A, Imai E *et al*: Bone marrow is a reservoir of repopulating mesangial cells during glomerular remodeling. *J. Am. Soc. Nephrol.* 12:2625-2635, 2001.

[6]   Rookmaaker MB, Smits AM, Tolboom H *et al*: Bone-marrow-derived cells contribute to glomerular endothelial repair in experimental glomerulonephritis. *Am. J. Pathol.* 163:553-562, 2003.

[7]   Akagi Y, Isaka Y, Akagi A *et al*: Transcriptional activation of a hybrid promoter composed of cytomegalovirus enhancer and beta-actin/beta-globin gene in glomerular epithelial cells in vivo. *Kidney Int.* 51:1265-1269, 1997.

[8]   Duffield JS, Park K, Hsaio L, Kelley V, Bonventre J: Tubular cell replenishment is independent of bone marrow stem cells (BMSCs) in the post-ischemic mouse kidney. (abstract) *J. Am. Soc. Nephrol.* 15:38A, 2004

[9] Duffield JS, Park KM, Hsiao LL *et al*: Restoration of tubular epithelial cells during repair of the postischemic kidney occurs independently of bone marrow-derived stem cells. *J Clin Invest* 115:1743-1755, 2005.

[10] Fang TC, Alison MR, Cook HT *et al*: Proliferation of Bone Marrow-Derived Cells Contributes to Regeneration after Folic Acid-Induced Acute Tubular Injury. *J. Am. Soc. Nephrol.* 16:1723-1732, 2005.

[11] Krause DS, Theise ND, Collector MI *et al*: Multi-organ, multi-lineage engraftment by a single bone marrow-derived stem cell. *Cell* 105:369-377, 2001.

[12] Szczypka MS, Westover AJ, Clouthier SG *et al*: Rare incorporation of bone marrow-derived cells into kidney after folic Acid-induced injury. *Stem Cells* 23:44-54, 2005.

[13] Gupta S, Verfaillie C, Chmielewski D *et al*: A role for extrarenal cells in the regeneration following acute renal failure. *Kidney Int.* 62:1285-1290, 2002.

[14] Lagaaij EL, Cramer-Knijnenburg GF, van Kemenade FJ *et al*: Endothelial cell chimerism after renal transplantation and vascular rejection. *Lancet* 357:33-37, 2001.

[15] Grimm PC, Nickerson P, Jeffery J *et al*: Neointimal and tubulointerstitial infiltration by recipient mesenchymal cells in chronic renal-allograft rejection. *N. Engl. J. Med.* 345:93-97, 2001.

[16] Nishida M, Kawakatsu H, Shiraishi I *et al*: Renal tubular regeneration by bone marrow-derived cells in a girl after bone marrow transplantation. *Am. J. Kidney Dis.* 42:E10-E12, 2003.

[17] Direkze NC, Forbes SJ, Brittan M *et al*: Multiple organ engraftment by bone-marrow-derived myofibroblasts and fibroblasts in bone-marrow-transplanted mice. *Stem Cells* 21:514-520, 2003.

[18] Rhim JA, Sandgren EP, Degen JL *et al*: Replacement of diseased mouse liver by hepatic cell transplantation. *Science* 263:1149-1152, 1994.

[19] Stevens ME, Meneses JJ, Pedersen RA: Expression of a mouse metallothionein-Escherichia coli beta-galactosidase fusion gene (MT-beta gal) in early mouse embryos. *Exp. Cell Res.* 183:319-325, 1989.

[20] Nilsson E, Lendahl U: Transient expression of a human beta-actin promoter/lacZ gene introduced into mouse embryos correlates with a low degree of methylation. *Mol. Reprod. Dev.* 34:149-157, 1993.

[21] Beddington RS, Morgernstern J, Land H, Hogan A: An in situ transgenic enzyme marker for the midgestation mouse embryo and the visualization of inner cell mass clones during early organogenesis. *Development* 106:37-46, 1989.

[22] Sands AT, Hansen TN, Demayo FJ *et al*: Cytoplasmic beta-actin promoter produces germ cell and preimplantation embryonic transgene expression. *Mol. Reprod. Dev.* 34:117-126, 1993.

[23] Schorpp M, Jager R, Schellander K *et al*: The human ubiquitin C promoter directs high ubiquitous expression of transgenes in mice. *Nucleic. Acids Res.* 24:1787-1788, 1996.

[24] Takeda S, Toyoda Y: Expression of SV40-lacZ gene in mouse preimplantation embryos after pronuclear microinjection. *Mol. Reprod. Dev.* 30:90-94, 1991.

[25] Boshart M, Weber F, Jahn G *et al*: A very strong enhancer is located upstream of an immediate early gene of human cytomegalovirus. *Cell* 41:521-530, 1985.

[26] Kisseberth WC, Brettingen NT, Lohse JK, Sandgren EP: Ubiquitous expression of marker transgenes in mice and rats. *Dev. Biol.* 214:128-138, 1999.

[27] Zambrowicz BP, Imamoto A, Fiering S *et al*: Disruption of overlapping transcripts in the ROSA beta geo 26 gene trap strain leads to widespread expression of beta-galactosidase in mouse embryos and hematopoietic cells. *Proc. Natl. Acad. Sci. USA* 94:3789-3794, 1997.

[28] Ahern-Rindell AJ, Prieur DJ, Murnane RD *et al*: Inherited lysosomal storage disease associated with deficiencies of beta-galactosidase and alpha-neuraminidase in sheep. *Am. J. Med. Genet* 31:39-56, 1988.

[29] Weiss DJ, Liggitt D, Clark JG: Histochemical discrimination of endogenous mammalian beta-galactosidase activity from that resulting from lac-Z gene expression. *Histochem J.* 31:231-236, 1999.

[30] Conchie J, Findlay J, Levvy GA: Mammalian glycosidases; distribution in the body. *Biochem. J.* 71:318-325, 1959.

[31] Pearson B, Wolf PL, Vazquez J: A comparative study of a series of new indolyl compounds to localize beta-galactosidase in tissues. *Lab. Invest.* 12:1249-1259, 1963.

[32] Cohen RB, Tsou KC, Rutenburg SH, Seligman AM: The colorimetric estimation and histochemical demonstration of beta-d-galactosidase. *J. Biol. Chem.* 195:239-249, 1952.

[33] Lojda Z: Indigogenic methods for glycosidases. II. An improved method for beta-D-galactosidase and its application to localization studies of the enzymes in the intestine and in other tissues. *Histochemie* 23:266-288, 1970.

[34] Lederberg J: The beta-d-galactosidase of Escherichia coli, strain K-12. *J. Bacteriol.* 60:381-392, 1950.

[35] Roufosse C, Bou-Gharios G, Prodromidi E *et al*: Bone Marrow-Derived Cells Do Not Contribute Significantly to Collagen I Synthesis in a Murine Model of Renal Fibrosis. *J. Am. Soc. Nephrol.*, 2006.

[36] Lee BY, Han JA, Im JS *et al*: Senescence-associated beta-galactosidase is lysosomal beta-galactosidase. *Aging Cell* 5:187-195, 2006.

[37] Ikawa M, Yamada S, Nakanishi T, Okabe M: 'Green mice' and their potential usage in biological research. *FEBS Lett.* 430:83-87, 1998.

[38] Okabe M, Ikawa M, Kominami K *et al*: 'Green mice' as a source of ubiquitous green cells. *FEBS Lett.* 407:313-319, 1997.

[39] Herrera MB, Bussolati B, Bruno S *et al*: Mesenchymal stem cells contribute to the renal repair of acute tubular epithelial injury. *Int. J. Mol. Med.* 14:1035-1041, 2004.

[40] Imasawa T, Utsunomiya Y, Kawamura T *et al*: The potential of bone marrow-derived cells to differentiate to glomerular mesangial cells. *J. Am. Soc. Nephrol.* 12:1401-1409, 2001.

[41] Iwasaki M, Adachi Y, Minamino K *et al*: Mobilization of bone marrow cells by G-CSF rescues mice from cisplatin-induced renal failure, and M-CSF enhances the effects of G-CSF. *J. Am. Soc. Nephrol.* 16:658-666, 2005.

[42] Lin F, Moran A, Igarashi P: Intrarenal cells, not bone marrow-derived cells, are the major source for regeneration in postischemic kidney. *J. Clin. Invest.* 115:1756-1764, 2005.

[43] Stokman G, Leemans JC, Claessen N *et al*: Hematopoietic stem cell mobilization therapy accelerates recovery of renal function independent of stem cell contribution. *J. Am. Soc. Nephrol.* 16:1684-1692, 2005.

[44] Cormack B, Valdivia R, Falkow S: FACS-optimized mutants of the green fluorescent protein (GFP). *Gene* 173:33-38, 1996.

[45] Ellenberg J, Lippincott-Schwartz J, Presley JF: Dual-colour imaging with GFP variants. *Trends Cell Biol.* 9:52-56, 1999.

[46] Ikarashi K, Li B, Suwa M *et al*: Bone marrow cells contribute to regeneration of damaged glomerular endothelial cells. *Kidney Int.* 67:1925-1933, 2005.

[47] Li B, Morioka T, Uchiyama M, Oite T: Bone marrow cell infusion ameliorates progressive glomerulosclerosis in an experimental rat model. *Kidney Int.* 69:323-330, 2006.

[48] Neale FC, Clubb JS, Hotchkis D, Posen S: Heat Stability of human placental alkaline phosphatase. *J Clin Pathol* 18:359-363, 1965

[49] Posen S, Cornish CJ, Horne M, Saini PK: Placental alkaline phosphatase and pregnancy. *Ann N Y Acad Sci* 166:733-744, 1969

[50] Broekema M, Harmsen MC, Koerts JA *et al*: Determinants of tubular bone marrow-derived cell engraftment after renal ischemia/reperfusion in rats. *Kidney Int.* 68:2572-2581, 2005.

[51] Broekema M, Harmsen MC, van Luyn MJ *et al*: Bone Marrow-Derived Myofibroblasts Contribute to the Renal Interstitial Myofibroblast Population and Produce Procollagen I after Ischemia/Reperfusion in Rats. *J. Am. Soc. Nephrol.* 18:165-175, 2007.

[52] Mujtaba T, Han SS, Fischer I *et al*: Stable expression of the alkaline phosphatase marker gene by neural cells in culture and after transplantation into the CNS using cells derived from a transgenic rat. *Exp. Neurol.* 174:48-57, 2002.

[53] Geusz ME, Fletcher C, Block GD *et al*: Long-term monitoring of circadian rhythms in c-fos gene expression from suprachiasmatic nucleus cultures. *Curr. Biol.* 7:758-766, 1997.

[54] Hastings JW: Chemistries and colors of bioluminescent reactions: a review. *Gene* 173:5-11, 1996

[55] Naylor LH: Reporter gene technology: the future looks bright. *Biochem. Pharmacol.* 58:749-757, 1999.

[56] Contag PR: Whole-animal cellular and molecular imaging to accelerate drug development. *Drug Discov. Today* 7:555-562, 2002.

[57] Greer LF, III, Szalay AA: Imaging of light emission from the expression of luciferases in living cells and organisms: a review. *Luminescence* 17:43-74, 2002.

[58] Cao YA, Bachmann MH, Beilhack A *et al*: Molecular imaging using labeled donor tissues reveals patterns of engraftment, rejection, and survival in transplantation. *Transplantation* 80:134-139, 2005.

[59] Imai E, Akagi Y, Isaka Y *et al*: Glowing podocytes in living mouse: transgenic mouse carrying a podocyte-specific promoter. *Exp. Nephrol.* 7:63-66, 1999.

[60] Iwano M, Plieth D, Danoff TM *et al*: Evidence that fibroblasts derive from epithelium during tissue fibrosis. *J. Clin. Invest* 110:341-350, 2002.

[61] Asahara T, Masuda H, Takahashi T *et al*: Bone marrow origin of endothelial progenitor cells responsible for postnatal vasculogenesis in physiological and pathological neovascularization. *Circ. Res.* 85:221-228, 1999.

[62] Finnegan EJ, Wang M, Waterhouse P: Gene silencing: fleshing out the bones. *Curr Biol* 11:R99-R102, 2001

[63] Pannell D, Ellis J: Silencing of gene expression: implications for design of retrovirus vectors. *Rev. Med. Virol.* 11:205-217, 2001.

[64] Togel F, Hu Z, Weiss K *et al*: Administered mesenchymal stem cells protect against ischemic acute renal failure through differentiation-independent mechanisms. *Am. J. Physiol. Renal. Physiol.*, 2005.

[65] Lange C, Togel F, Ittrich H *et al*: Administered mesenchymal stem cells enhance recovery from ischemia/reperfusion-induced acute renal failure in rats. *Kidney Int.* 68:1613-1617, 2005.

[66] Karrer FM, Reitz BL, Hao L, Lafferty KJ: Fluorescein labeling of murine hepatocytes for identification after intrahepatic transplantation. *Transplant Proc.* 24:2820-2821, 1992.

In: Developments in Stem Cell Research
Editor: Prasad S. Koka

ISBN: 978-1-60456-341-2
© 2008 Nova Science Publishers, Inc.

Chapter 8

THE REGULATORY FUNCTION OF MICRORNA IN STEM CELLS

Wei Wu[*1]*, Vivien Wang*[2] *and*
Gang-Ming Zou[*3]

[1]Division of Cancer Biology, Department of Medicine, Northwestern University,
Evanston, IL 60201, USA
[2]Department of Pathology, Evanston Northwestern Hospital, Evanston, IL 60201, USA
[3]Department of Pathology, John Hopkins University School of Medicine, 1550 Orleans
Street. CRB-2, M341, Baltimore, MD 21231, USA

ABSTRACT

MicroRNAs (miRNAs) are endogenous ~22 nucleotide non-coding RNAs. They play important regulatory roles in plants and in animals by pairing to messenger RNAs of target genes, specifying mRNA cleavage or repression of protein synthesis. Recent evidences indicate that they exhibit important regulatory roles in development timing, cell proliferation, cell survival and apoptosis. miRNAs regulate normal stem cell development both in mammalian and non-mammalian systems. For instance, they regulate stem cell self-renewal and differentiation, and have gained increasing attention in stem cell biology. Here we review recent progression of miRNAs in stem cells research, and summarize current understanding of miRNA in their expression in stem cells and their cellular biological role in stem cells.

Keywords: *microRNA, miRNA, stem cell, ES cells, cell cycle, c-kit, differentiation.*

[*] **Corresponding authors**: Wei Wu, Division of Cancer Biology, Department of Medicine, Northwestern University, ENH research Institute, 1001 University Place, Evanston, IL 60201, USA. Telephone: 1-224-364-7514. Fax: 1-224-364-7402; Email:wei-wu-0@northwestern.edu or Gang-Ming Zou. Department of Pathology, John Hopkins University School of Medicine, 1550 Orleans Street. CRB-2, M341, Baltimore, MD 21231, USA Tel: 1-410-955 3511. Fax: 1-410-614 0671. Email: gzou1@jhmi.edu

Stem cells are a term to define a cell that can differentiate into multiple cell type and maintain self-renewal activity. They can be classified to three major groups: embryonic stem (ES) cells, embryonic germ cells, and adult stem cells. ES cells are cells derived from the inner cell mass of developing blastocyst. These cells are self-renewaling, pluripotent, and theoretically immortal. They can differentiate into three germlayers: ectoderm, mesoderm, and endoderm. Embryonic germ cells are collected from the fetus later in the developmental process from the gonadal ridge. The characterstics of embryonic germ cells are similar to ES cells. These cells can also give rise to the three germ layers whereas the cell type that develop from embryonic germ cells are slightly limited than those that develop from ES cells. Adult stem cells are found in different tissue of the developed adult organ that remain in an undifferentiated or unspecialized state. They can give rise to specialized cell type of the tissue from which they came. Stem cells have three distinctive properties: self-renewal, the capability to develop into multiple lineages and the potential to proliferate [1]. They exist and persist in many tissues or organs such as brain, blood, bone marrow, breast, etc. Transcription factors, such as Oct-4, SOX2, and NAOG, are essential for human embryonic stem cells self-renewal [2]. Small non-coding RNAs have emerged as potent regulators of gene expression at both transcriptional and post-transcriptional levels in diverse organisms. Two major classes of small RNAs are small interfering RNAs, microRNA (miRNAs), which are of ~21-25 nucleotides (nt) in length, negatively regulate the expression of protein encoding genes. The PIWI/Argonaute interacting RNAs (piRNAs) are newly discovered small non-coding RNAs in mouse genome, which are ~30 nt in length [3]. Recent studies have demonstrated the importance of miRNAs in embryonic stem cell differentiation, limb development, adipogenesis, myogenesis, angiogenesis and hematopoiesis, neurogenesis, and epithelial morphogenesis [4]. piRNA interacts with MIWI , a murine PIWI family member. It regulates mouse germline stem cell developments [5]. Here we mainly discuss the miRNAs expression profile and their signaling pathways in stem cell regulations.

MicroRNA Biogenesis and Its Functions

The first miRNA, lin-4, was initially discovered over a decade ago in Caenorhabditis elegans [6, 7]. Because of the versatile functions, miRNAs were widely recognized in 2001 [8-10]. Currently, miRNAs have been identified in a wide array of organisms, including plants, zebrafish, Drosophila, and mammals [11]. The expression of miRNAs in multicellular organisms exhibits spatiotemporal, tissue- and cell-specificity, suggesting their involvement in tissue morphogenesis and cell differentiation [12]. To date, the public miRNA database, MIRBase, has collected 462 human and 340 mouse miRNA sequences (http://microrna.sanger.ac.uk; release 8.1).

miRNA biogenesis is a comprehensive biochemical process [13]. Briefly, miRNAs are transcribed by RNA polymerase II and initially generated by a large RNA precursor---a primary transcript RNA (pri-miRNA). The pri-miRNA precessed into a stem-loop structure of about 70-100 nt (pre-miRNA) by a double-strand RNA-specific ribonuclease, Drosha, and its interaction partner DGCR8/pasha [14, 15]. These pre-miRNAs are transported into the cytoplasm via an Exportin-5-RanGTP dependent mechanism [16]. In the cytoplasm, the pre-miRNAs are cleaved by a second, dsRNA-specific ribonuclease called Dicer with the help of

TRBP and AGO2 [17]. In *Drosophila*, there are two distinct Dicers (Dcr1/2). Dcr1 specializes in processing endogenous hairpin RNA precursors into miRNAs, and Dcr2 specializes in the cleavage of double stranded RNAs (dsRNAs) that are destined to function as siRNAs [18]. Loss of dicer-1 completely disrupts the miRNA pathway and only has a weak effect on the siRNA pathway. Dicer-1 also interacts with Loquacious, the double-stranded RNA-binding domain protein to mediate pre-miRNA process [19]. The Loquacious appears to be required for normal miRNA maturation and germ line stem cell maintenance [20].

The mature miRNA (17-25 nt) is bound by a complex called miRNA-associated RNA-induced silencing complex (miRISC). This complex binds to 3' UTR of target mRNA, either degrades mRNA or represses translation of mRNA depending on the degree of complementarity. Most animal miRNAs act as inhibitor in translation of mRNA; however, miRNA-directed mRNA cleavage has also been shown to occur in mammals [21-23]. Therefore, miRNA may share with siRNA pathway in certain cell content.

MicroRNA Expression Profile in Stem Cells

Several genes, including transcription factors such as Oct-4 [24], was thought to be essential players in embryonic stem cell self-renewal . Several studies identified unique profilings of miRNAs in stem cells [2], suggesting that miRNAs may play a critical role in the maintenance of the pluripotent cell state and in the regulation of early mammalian development. For instance, Houbaviy *et al* [25] identified the different miRNA profile between undifferentiated and differentiated mouse embryonic stem (mES) cells. The ES cell specific miRNAs are encoded by genomic loci clustered within 2.2 kb. The expression of ES specific miRNAs are repressed when ES cells differentiate into embryoid bodies. In contrast, the levels of many previously described miRNAs remain constant or increase upon differentiation. Recently, the miRNA expression in single ES cells has been developed on the basis of RT-PCR techniques [26]. Suh *et al*. [27] subsequently utilized the human embryonic stem cell system to clone the miRNAs and dissect 16 known miRNAs, 17 novel miRNAs and 3 (miR296, miR-301 and miR-302) compared with mES. miR-371-372-372*-373 cluster is down-regulated during differentiation in human ES cells. In an *in vivo* study, Dicer-1-deficient mice die very early in around embryonic day 7.5, with essentially a complete loss of pluripotent stem cells [28]. Taken together, these studies indicate programmed expression of miRNAs in stem cell, coupled with coding gene expression which is important for stem cell functioning.

miRNA Expression in Hematopoietic Stem Cells and Progenitors

Chen *et al* [29] have cloned about 100 unique miRNAs from mouse bone marrow. They have identified miR-223 as a mouse bone marrow specific miRNA, which is expressed mainly in myeloid lineage. miR-342 is a mouse spleen-specific miRNA which is expressed mainly in T and B cells [30]. Ectopically expression of hematopoietic miRNAs in hematopoietic stem/progenitor cells reveals that miR-181a is preferentially expressed in the B cells in mouse bone marrow. Ectopic expression MiR-181a in hematopoietic stem/progenitor cells promotes B-cell differentiation [30]. miRNA is also involved in erythroid differentiation of $CD34^+$ cord blood hematopoietic progenitor cells. The level of miR-221 and miR-222 were down-regulated in erythroid progenitor cells compared with $CD34^+$ cord blood hematopoietic stem cells. Furthermore, transfection of $CD34^+$ progenitor cells with miR-221 and miR-222 caused impaired proliferation and accelerated differentiation into erythropoietic cells, [31]. In human myeloid progenitor cells, miR-223 expression is low whereas its expression is upregulated following retinoic acid –induced differentiation. Furthermore, knockdown of miR-223 impaired the differentiation response to inducer. These studies demonstrate miR-223 plays a crucial function during granulopoiesis [32]. It seems like that different lineage of HSC exhibits specific miRNA marker, which may be used to define the progenitor cells, but more data is needed to support this hypothesis.

MicroRNA Regulate Cell Cycles and Stem Cell Division

Cell cycle regulators control the stem cell self-renewal and determine the cell fates, miRNAs may regulate G1/S transition in stem cell division. Using the Drosophila germline stem cells (GSC) as a model, Hatfield *et al.* specifically inactivate the Dicer-1, which is required for miRNA pathway. They found that the rate of cell division in dcr-1 mutant GSCs was decreased. The cell cycle markers Cyclin E is high expressed, and the S phase is reduced. These studies indicate that perturbation of the miRNA pathway by mutant dcr-1 in GSCs delays the cell cycle at the G/S transition. Their studies further documented that miRNA may bind to 3'UTR of Daccpo gene (a homologue of the p21/p27 family of cyclin-dependent kinase inhibitors) and inhibit its activity; consequently, activate the Cyclin E protein . Murchison *et al* [33] recently generated Dicer deficient (exon 18, 22-23 in dicer gene) mouse ES cells ($DCR^{-/-}$), they observed these $DCR^{-/-}$ mouse ES cells exhibit a significant proliferative defect comparing to wild type or heterozygomous ES cells. This phenotype change is consistent with other dicer knockout (exons 18-20) mouse ES cells ($DCR^{\Delta/\Delta}$ ES) [34]. Moreover, the $DCR^{-/-}$ ES cells altered cell cycle profile which increase in G_1 and G_0 cells and a corresponding decrease of cells in G_2 and M phase.

Collectively, specific miRNAs turn on or off the gate of cell cycle, which regulate the stem cells fate under physiological conditions (Figure 2).

MicroRNA and Stem Cells Differentiation

miRNAs regulate cancer cell differentiation [30, 35], control over lineage determination in hematopoiesis [29], and neuronal stem cell differentiation [36]. miR-181 was preferentially expressed in the B-lymphoid cells of mouse bone marrow. Its ectopic expression in hematopoietic stem/progenitor cells led to an increased differentiation of B-lineage cells. Kanellopoulou *et al.* [34] recently observed that $DCR^{\Delta/\Delta}$ ES cells express similar level of Oct-4 compared with DCR wildtype ES cells. These $DCR^{\Delta/\Delta}$ ES cell also expressed the ES cell-specific pre-miRNAs, miR-292a and miRNA-293. However, when 12 days embroid bodies

(Ebs) were examined, the DCR$^{\Delta/\Delta}$ cells with stem-cell properties (high Oct-4 level) only formed aggregates of cells with little morphological evidence of differentiation. Currently, the question which miRNA is responsible for the ES-derived differentiation is not clear. Using mouse ES cell-derived neurogenesis model and miRNA array technology, Krichevsky *et al.* found that most of miRNAs were up-regulated during neuron development; miR-124a and miR-9 were strongly abundant in differentiated neuron cells [36]. Both gain of function and loss of function experiments support the role of these miRNAs in regulation of stem cell differentiation. Therefore, understanding the differentiation specific miRNAs would aid in developing potential clinical application of specific regulator of miRNA expression.

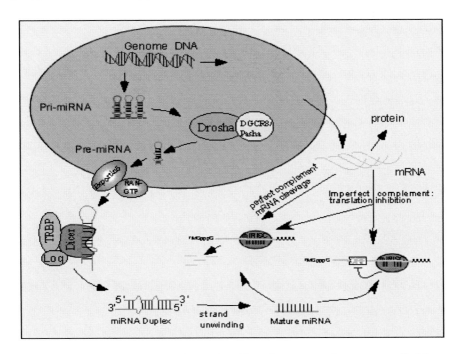

Figure 1. Diagram of miRNA biogenesis and its action. The primary transcripts are transcribed from genomic DNA, then processed by the RNase III enzyme Drosha and its co-factors, DGCR8/Pasha, to generate the pre-miRNA products (~70 nt). The pre-miRNA is exported from the nucleus to the cytoplasm by Exportin 5/Ran-GTP. Subsequently, a second RNasIII enzyme, Dicer, together with its partner, such as the trans-activator RNA-binding protein (TRBP), or Loquacious (Loq), produces miRNA duplexes (~22 nt). Mature miRNA is produced, and guarded by the effector complex, named miRNA-associated RNA-induced silencing complex (miRISC) to target mRNAs. miRNAs regulate the target genes by translational inhibition or mRNA degradation.

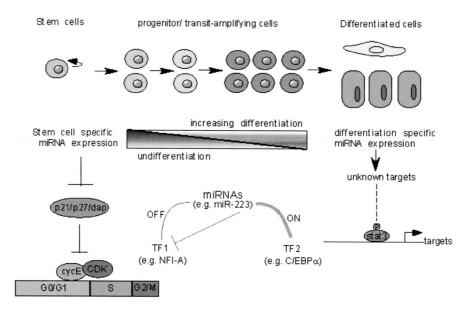

Figure 2. Model of miRNAs signaling in stem cell regulations. Stem Cell division acts through miRNA-dependent regulation of P21/P27 (left). During the differentiation, different transcription factors (TF) regulate miRNAs expresion in order and in turn. miRNAs expression inhibits the TF functions depending upon the cellular physiologic condition. For instance, Nuclear Factor 1 family of proteins (NF1-A), C/EBPα and miRNA-223 form the regulatory network to mediate stem cell differentiation. miRNAs may regulate the transcription factors such as STAT3 in directly or indirectly manner.

MicroRNA Mediated Signal Transduction Pathways in Stem Cells

miRNAs can form extensive regulatory networks with a complexity comparable to that of transcription factor [37]. miRNA pathways have been defined in timing development in C. elegans such as lin4 and let 7. A recent study has shown that let-7 inhibits Ras activity in human lung cancer [38]. miRNA-mediated signal transduction pathways in stem cells has been proposed recently. Shcherbata *et al.* [39] hypothesized that the environmental regulation of Drosophila GSC division could act through miRNA-dependent regulation of P21/P27 /Dacapo. When the conditions are unfavorable for division, the key miRNAs are down-regulated, resulting in an increased p21/p27/Dacapo level, which subsequently halts the cell cycle in G1/S (Figure 2). We have suggested recently an optimal ratio between serine and tyrosine phosphorylation of STAT3 is a key condition to keep mouse ES cell pluripotency [40]. Activation of STAT3 also plays an important role in neuronal development, particularly in inhibiting neuronal terminal differentiation from neuronal stem cells [41, 42]. In mouse ES cell-derived neural precursors, miR-9 and miR-124a act in the STAT3 signaling pathway through phosphorylation of STAT3 protein without affecting the mRNA and protein level. MiR-9 inhibit STAT3 Tyr705 phosphorylation in those neural precursors [36] (Figure 2). Currently, there is no ample evidence showing miRNA posttranslationally regulated targets, one possible explanation might be indirect effect of miR-9 on STAT3. Both miR-221 and miR-222 affect c-Kit expression in human umbilical vein endothelial cells ; as a consequence, they affect the angiogenic properties of stem cell factor, the ligand of c-Kit. Interaction between miR-222 and c-Kit is likely to be part of a complex circuit that controls the ability of endothelial cells to form new capillaries [43]. Since SCF-Kit signaling pathway plays an important role as a survival factor for many type of stem/progenitor cells, including hematopoietic stem cell and neuronal stem cells [44], one can speculate that miRNA regulate stem cell survival via regulation of c-Kit signaling. miR-223 was identified as an important modulator of human myeloid differentiation. Interestingly, NF1-A inhibits miR-223 expression in undifferentiated progenitor cells, whereas c/EBPα binds the promoter of miR-223 in replacement of NF1-A and stimulates miR-223 expression during differentiation [32]. Gene regulation in stem cells are comprehensive, the regulatory circuitry between miRNAs and target genes appears crucial, however, new miRNA pathways remain to be uncovered in the future.

CONCLUSION

The miRNA field is rapidly developing with the members of miRNA family. Increasing number of miRNAs have been identified and their function is under investigation. Recent findings suggest that miRNA plays an important role in controlling cell viability, cell cycle, proliferation, or apoptosis. Exploration of miRNA functions in stem cells may help us to understand self-renewal and differentiation of stem cells in integrity genome. Continued research into miRNA functions might unveil a new molecular taxonomy of stem cell. Furthermore, elucidation of how miRNAs and other small RNAs such as piRNA interact with known and yet-to-be identified gene regulatory pathways in different stem cells should provide us with a more in-depth understanding of the mechanisms in regulation of cellular function of stem cells, and facilitate the application of small RNA technology in stem cell related disease control and treatment.

ACKNOWLEDGEMENTS

We apologize that we were able to cite only a limited number of references due to the space limitation and the focus of this article.

REFERENCES

[1] Jordan CT, Guzman ML, Noble M. Cancer stem cells. *N. Engl. J. Med.* 2006; 355:1253-1261.

[2] Zhang B, Pan X, Anderson TA. MicroRNA: a new player in stem cells. *J. Cell Physiol.* 2006; 209:266-269.

[3] Grivna ST, Beyret E, Wang Z, Lin H. A novel class of small RNAs in mouse spermatogenic cells. *Genes Dev.* 2006; 20:1709-1714.

[4] Hornstein E, Mansfield JH, Yekta S, Hu JK, Harfe BD, McManus MT, et al. The microRNA miR-196 acts upstream of Hoxb8 and Shh in limb development. *Nature* 2005; 438:671-674.

[5] Grivna ST, Pyhtila B, Lin H. MIWI associates with translational machinery and PIWI-interacting RNAs (piRNAs) in regulating spermatogenesis. *Proc. Natl. Acad. Sci. USA* 2006;103:13415-13420.

[6] Lee RC, Feinbaum RL, Ambros V. The C. elegans heterochronic gene lin-4 encodes small RNAs with antisense complementarity to lin-14. *Cell* 1993; 75:843-854.

[7] Wightman B, Ha I, Ruvkun G. Posttranscriptional regulation of the heterochronic gene lin-14 by lin-4 mediates temporal pattern formation in C. elegans. *Cell* 1993; 75:855-862.

[8] Lagos-Quintana M, Rauhut R, Lendeckel W, Tuschl T. Identification of novel genes coding for small expressed RNAs. *Science* 2001; 294:853-858.

[9] Lau NC, Lim LP, Weinstein EG, Bartel DP. An abundant class of tiny RNAs with probable regulatory roles in Caenorhabditis elegans. *Science* 2001; 294:858-862.

[10] Lee RC, Ambros V. An extensive class of small RNAs in Caenorhabditis elegans. *Science* 2001; 294:862-864.

[11] Kim VN, Nam JW. Genomics of microRNA. *Trends Genet* 2006; 22:165-173.

[12] Bartel DP. MicroRNAs: genomics, biogenesis, mechanism, and function. *Cell* 2004; 116:281-297.

[13] Esquela-Kerscher A, Slack FJ. Oncomirs - microRNAs with a role in cancer. *Nat. Rev. Cancer* 2006; 6:259-269.

[14] Lee Y, Jeon K, Lee JT, Kim S, Kim VN. MicroRNA maturation: stepwise processing and subcellular localization. *Embo J.* 2002; 21:4663-4670.

[15] Denli AM, Tops BB, Plasterk RH, Ketting RF, Hannon GJ. Processing of primary microRNAs by the Microprocessor complex. *Nature* 2004; 432:231-235.

[16] Lund E, Guttinger S, Calado A, Dahlberg JE, Kutay U. Nuclear export of microRNA precursors. *Science* 2004; 303:95-98.

[17] Chendrimada TP, Gregory RI, Kumaraswamy E, Norman J, Cooch N, Nishikura K, et al. TRBP recruits the Dicer complex to Ago2 for microRNA processing and gene silencing. *Nature* 2005; 436:740-744.

[18] Tomari Y, Zamore PD. Perspective: machines for RNAi. *Genes Dev.* 2005; 19:517-529.

[19] Saito K, Ishizuka A, Siomi H, Siomi MC. Processing of pre-microRNAs by the Dicer-1-Loquacious complex in Drosophila cells. *PLoS Biol* 2005; 3:e235.

[20] Forstemann K, Tomari Y, Du T, Vagin VV, Denli AM, Bratu DP, et al. Normal microRNA maturation and germ-line stem cell maintenance requires Loquacious, a double-stranded RNA-binding domain protein. *PLoS Biol.* 2005; 3:e236.

[21] Yekta S, Shih IH, Bartel DP. MicroRNA-directed cleavage of HOXB8 mRNA. *Science* 2004; 304:594-596.

[22] Bagga S, Bracht J, Hunter S, Massirer K, Holtz J, Eachus R, et al. Regulation by let-7 and lin-4 miRNAs results in target mRNA degradation. *Cell* 2005; 122:553-563.

[23] Lim LP, Lau NC, Garrett-Engele P, Grimson A, Schelter JM, Castle J, et al. Microarray analysis shows that some microRNAs downregulate large numbers of target mRNAs. *Nature* 2005; 433:769-773.

[24] Niwa H, Miyazaki J, Smith AG. Quantitative expression of Oct-3/4 defines differentiation, dedifferentiation or self-renewal of ES cells. *Nat Genet* 2000; 24:372-376.

[25] Houbaviy HB, Murray MF, Sharp PA. Embryonic stem cell-specific MicroRNAs. *Dev Cell* 2003; 5:351-358.

[26] Tang F, Hajkova P, Barton SC, Lao K, Surani MA. MicroRNA expression profiling of single whole embryonic stem cells. *Nucleic Acids Res.* 2006; 34:e9.

[27] Suh MR, Lee Y, Kim JY, Kim SK, Moon SH, Lee JY, et al. Human embryonic stem cells express a unique set of microRNAs. *Dev. Biol.* 2004; 270:488-498.

[28] Bernstein E, Kim SY, Carmell MA, Murchison EP, Alcorn H, Li MZ, et al. Dicer is essential for mouse development. *Nat. Genet* 2003; 35:215-217.

[29] Chen CZ, Li L, Lodish HF, Bartel DP. MicroRNAs modulate hematopoietic lineage differentiation. *Science* 2004; 303:83-86.

[30] Chen CZ, Lodish HF. MicroRNAs as regulators of mammalian hematopoiesis. *Semin. Immunol.* 2005; 17:155-165.

[31] Felli N, Fontana L, Pelosi E, Botta R, Bonci D, Facchiano F, et al. MicroRNAs 221 and 222 inhibit normal erythropoiesis and erythroleukemic cell growth via kit receptor down-modulation. *Proc. Natl. Acad. Sci. USA* 2005; 102:18081-18086.

[32] Fazi F, Rosa A, Fatica A, Gelmetti V, De Marchis ML, Nervi C, et al. A minicircuitry comprised of microRNA-223 and transcription factors NFI-A and C/EBPalpha regulates human granulopoiesis. *Cell* 2005; 123:819-831.

[33] Murchison EP, Partridge JF, Tam OH, Cheloufi S, Hannon GJ. Characterization of Dicer-deficient murine embryonic stem cells. *Proc. Natl. Acad. Sci. USA* 2005; 102:12135-12140.

[34] Kanellopoulou C, Muljo SA, Kung AL, Ganesan S, Drapkin R, Jenuwein T, et al. Dicer-deficient mouse embryonic stem cells are defective in differentiation and centromeric silencing. *Genes Dev.* 2005; 19:489-501.

[35] Sempere LF, Freemantle S, Pitha-Rowe I, Moss E, Dmitrovsky E, Ambros V. Expression profiling of mammalian microRNAs uncovers a subset of brain-expressed microRNAs with possible roles in murine and human neuronal differentiation. *Genome Biol.* 2004; 5:R13.

[36] Krichevsky AM, Sonntag KC, Isacson O, Kosik KS. Specific microRNAs modulate embryonic stem cell-derived neurogenesis. *Stem Cells* 2006; 24:857-864.

[37] Hobert O. Common logic of transcription factor and microRNA action. *Trends Biochem. Sci.* 2004; 29:462-468.

[38] Wu W, Sun M, Zou GM, Chen J. MicroRNA and cancer: Current status and prospective. *Int. J. Cancer* 2007; 120:953-960.

[39] Shcherbata HR, Hatfield S, Ward EJ, Reynolds S, Fischer KA, Ruohola-Baker H. The MicroRNA pathway plays a regulatory role in stem cell division. *Cell Cycle* 2006; 5:172-175.

[40] Zou GM, Chen JJ, Ni J. LIGHT induces differentiation of mouse embryonic stem cells associated with activation of ERK5. *Oncogene* 2006; 25:463-469.

[41] Moon C, Yoo JY, Matarazzo V, Sung YK, Kim EJ, Ronnett GV. Leukemia inhibitory factor inhibits neuronal terminal differentiation through STAT3 activation. *Proc. Natl. Acad. Sci. USA* 2002; 99:9015-9020.

[42] Gu F, Hata R, Ma YJ, Tanaka J, Mitsuda N, Kumon Y, et al. Suppression of Stat3 promotes neurogenesis in cultured neural stem cells. *J. Neurosci. Res.* 2005; 81:163-171.

[43] Poliseno L, Tuccoli A, Mariani L, Evangelista M, Citti L, Woods K, et al. MicroRNAs modulate the angiogenic properties of HUVECs. *Blood* 2006; 108:3068-3071.

[44] Zou GM. Protein tyrosine phosphatase Shp-2 in cytokine signalings and stem cell regulations. In "*Stem Cell Research Developments*". Calvin A. Fong (Ed). Nova Science Publisher. Hauppauge. NY. 2007. pp107-134.

In: Developments in Stem Cell Research
Editor: Prasad S. Koka

ISBN: 978-1-60456-341-2
© 2008 Nova Science Publishers, Inc.

Chapter 9

NICHE THEORY, STEM-STROMAL IMBALANCE AND APLASTIC ANAEMIA

*Sujata Law and Samaresh Chaudhuri**
Stem Cell Research and Application Unit
Department of Biochemistry and Medical Biotechnology, School of Tropical Medicine
C R Avenue, Kolkata-700073, India

ABSTRACT

Aplastic Anaemia is correlated with Stem Cell disorder, the detail mechanism of which still awaits extensive evaluation. Apart from "Inherited Aplastic Anaemia", the "Acquired Aplastic Anaemia", has also been considered to involve stem cell deficiency or disorder or more currently considered disorganized "stem cell niche". The manifestations in such types of Aplastic Anaemia are varied and supposed to impair the immunological functions as well. Causes for Acquired Aplastic Anaemia can also be multifactorial that may concern the stem cell itself or its microenvironment or the stem cell "niche", which is, however, in controversy. Studies conducted in our laboratory revealed that the stromal microenvironment is significantly affected in Acquired Aplastic Anaemia, especially in experimental animals and farmers exposed to insecticides/ pesticides of organophosphate and organochloride origin. Supplementation of healthy microenvironmental components including stromal cells and cord blood plasma factor (CBPF) have been found to rectify the stem cell deficiency under the event. It is suggested that Acquired Aplastic Anaemia can be recovered reversibly to health by supplementation of microenvironmental factors together with immune-reconstitution. This review attempts to re-establish the involvement of niche system and the role of

---

* **Corresponding Author:** Dr. S. Chaudhuri, Professor and Head, Department of Biochemistry and Medical Biotechnology, Stem Cell Research and Application Unit, Calcutta School of Tropical Medicine, C R Avenue, Kolkata-700073, India. E-mail:- samc2009@rediffmail.com; Phone:- 91-33-2241-4900/4065 Ext.224. Cell Phone: 9831-386835; Fax:- 91-33-22414915.

microenvironment /microenvironmental factors as therapeutic adjunct to the patient-concerned.

INTRODUCTION

Aplastic Anaemia is a disorder characterized by pancytopenia and bone marrow hypoplasia and is considered to be a bone marrow failure syndrome [1, 2]. Paul Ehrlich introduced the concept of Aplastic Anaemia in 1888 when he studied a case of a pregnant woman who died of bone marrow failure [3]. However, it was not until 1904 that Chauffard named this disorder as Aplastic Anaemia [4]. In most patients with Acquired Aplastic Anaemia, bone marrow failure is believed to result from immunologically mediated destruction of the haematopoietic progenitor cells (HPC) within the marrow [5, 6]. Although, there can be reasons manifold, the immunological abnormalities are considered to play a prominent role in bone marrow (BM) aplasia. Activated T-Cells may directly kill haematopoietic progenitor cells and release haematopoietic inhibitors such as interferon (IFN) and Tumor Necrosis Factor (TNF).

A number of recent studies revealed that the basis of bone marrow failure is seeded with primary defects in or damage to the stem cell or the marrow microenvironment or other functional discrepancies [7-16]. The distinction between acquired and inherited disease may present clinical challenge but more than 80% of the cases has been found to be acquired. However, other reports more logistically support the component damage or abnormality within the bone marrow niche [17].

On morphological evaluation, the bone marrow is found to be deficient in haematopoietic elements, showing largely fat cells. Studies with immuno-cytochemistry, flowcytometry (FACS) and other showed that the CD34$^+$ cell population, which contains stem cells and the early committed progenitors, is substantially reduced [13,18]. Controversy arises whether this is due to an impaired generation of the pluripotent cells within the bone marrow or is the cytotoxic effect of the activated T cells in course of autoimmune induction. Data from in-vitro colony culture assays or those from immunological assays suggest profound functional loss of the haematopoietic progenitors, so much so that they are unresponsive to high levels of haematopoietic growth factors [19-26].

STEM - STROMAL IMBALANCE

A collosum amount of literature have been accumulated concerning the role of marrow microenvironment for maintaining the proper "stemness" within the bone marrow [27]. Current studies in our laboratory also furnished that the microenvironmental factors can stimulate the stem cell proliferation and function [28]. These observations simultaneously support the view that in Aplastic Anaemia the microenvironmental support is also disorganized [29]. In experimentally induced Aplastic Anaemia, such observations are pronounced with the fact that stromal marrow microenvironment suffer a huge cellular disorganization and henceforth affecting the healthy stem cell generation and differentiation [30-32]. It remains to be evaluated properly whether all the components of the bone marrow microenvironment or the niche are affected in a chain reaction or due to partial loss of continuity [33, 34]. It is more likely that the stem stromal relationship forms the basis of healthy stem cell differentiation and proliferation. The mesenchymal stem cells have the peculiar property of reversible and transient immortalization in association with primary fibroblast like CD34 negative stem cells [35, 36].

The realization that stable-custom microenvironment might contain haematopoietic stem cells to all such regions called "niches"[37], helped a lot in understanding of stem cell generation, its nourishment, and functional propagation in the presence of subsets of tissue cells and extracellular substances, most possibly the cytokines and growth factors [38-41]. This is considered to control their self-renewal and progeny production *in-vivo* [42]. We review, hereby, our current understanding to disorganization of this stable custom marrow microenvironment or in other term "the niche" that may lead to interruption of stem cell renewal with a consequent deficient or "failure bone marrow". Thus the hypothesis that bone marrow failure is a lesion of "soil" rather than "seed" [16] was based on several experimental observations. There can be various defects in the bone marrow niches due to microenvironmental disbalance leading to atrophy of so called niches. Histologically, these showed atrophy of the sinusoidal vasculature, abnormal adipocyte proliferation, intrinsic differences between the red and yellow marrow and finally the stromal cell distribution [43, 44]. Direct evidence of bone marrow failure due to stromal defect has been observed in genetically defective mouse strain (Sl / Sl^d) which was not corrected by stem cell infusion [45-47]. This indicated a clear interdependence of stem-stromal relationship, which is bridged by various components including cytokines and growth factors and many more.

Further evidences showed that bone marrow stromal cell defect did not clearly correlate the pathogenesis of aplastic anaemia. The question evolved at the event is whether the stem stromal access is the vulnerable region that may be affected in Aplastic Anaemia. The Acquired Aplastic Anaemia can be an ideal model to study such defects and in our laboratory such conditions have been induced in mice with prolonged exposures of pesticides in combination. The components of the haemopoietic niches within the bone marrow thus may act in unknown mysterious manner, the sequel of which is still under investigation. Our experience suggests that the unique stem-stromal relationship is deregulated at the event of Acquired Aplastic Anaemia in animal system as was induced by prolonged pesticide exposure and this corresponds clearly with the disorganized niche structure under the event (Figure 1 a, b, c and 2 a, b, c).

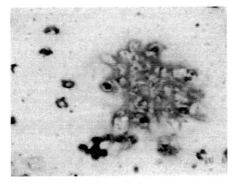

Figure 1a. "Niche" structure from bone marrow smear of normal mice representing the matrix pact with stromal fibroblasts and the pluripotent cells; other structure related to microenvironment are also seen.

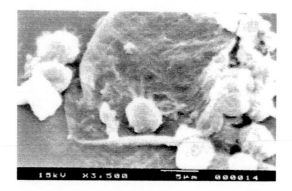

Figure 1b. Scanning Electron Microscopic (SEM) structure of a designated "niche" from the bone marrow of normal mice (N): shows the matrix together with stromal cord and the generating pluripotent stem cells. Associated microenvironmental structures are also seen.

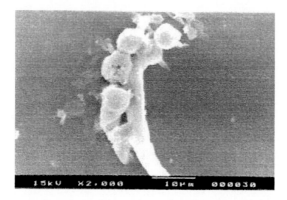

Figure 1c. SEM picture of stromal cord and the released stem cells from the bone marrow of normal mice.

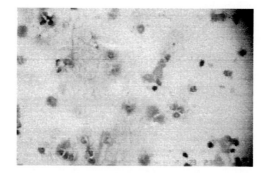

Figure 2a. "Niche" structure from bone marrow smear of Aplastic Anaemia induced mice: shows distorted and abnormal configuration with scanty pluripotent cell distribution. The associated microenvironmental components are also absent.

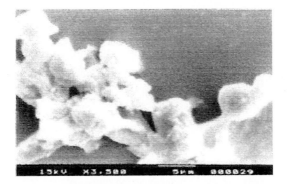

Figure 2b. The "niche" structure under SEM from the bone marrow of Aplastic Anaemia induced mice: shows disorganized "niche" structure with tortuous matrix, the environmental components are absent.

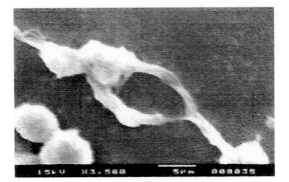

Figure 2c. The SEM feature of stromal cord from the bone marrow of Aplastic Anaemia induced mice. This shows the distorted stromal cord with disorganized stem cell release system. The pluripotent stem cell structures are also distorted.

PESTICIDE INDUCED ACQUIRED APLASTIC ANAEMIA IN HUMAN

Evidences of Acquired Aplastic Anaemia in human have been reported earlier [48, 49]. These showed that various agents can be responsible for inducing this secondary Aplastic Anaemia such as radiation, drugs and chemicals including organophosphates and organochlorides, metals, viruses, immunological diseases and many other related factors. Apart from this, idiopathic Aplastic Anaemia has also been accounted for, which also represented bone marrow disorganization. Prolonged use of antibiotics, chemotherapeutic agents, for examples, cytotoxic drugs, have shown a significant number of Aplastic Anaemia cases with proven bone marrow failure [50-52].

The deleterious effects of pesticides to the farmers with particular reference to haematologic considerations have been discussed by Issaragrisil et al ,2006, [53] in which he showed important association with Aplastic Anaemia in rural Thailand with individuals of farming practices. Significant association has been observed with agricultural pesticides, namely organophosphates, DDT, carbamates etc.; remarkable risk estimate has been established with organophosphate (2.1) DDT (6.7) and carbamates (7.4). In India particularly in rural West Bengal we have observed significant incidence of Aplastic Anaemia and Hypoplastic anaemia amongst farmers having regular practice of pesticide usage. Eventually in our hospital we received a statistically significant number of Acquired Aplastic Anaemia patient who had prolonged exposure of pesticides in course of agricultural work. Issaragrisil et al in his remarkable reporting has rightly mentioned that the incidence of Acquired Aplastic Anaemia is much greater in Asia compared to that found in Europe and United States. This basic concept of pesticide hazard has stimulated the controlled use of different types of pesticides including DDT, which had a wide use in the Western countries. Interesting reports are also available in this context that was reviewed by Young, 2002, [54] and Muir et al, 2003 [55]. Surveys involving insecticide exposure to human directed towards Aplastic Anaemia, accounted for 2-9% of the disease incidence in USA and some countries of Europe [56-59]. Even the use of household pesticide for a continuous period has also reported incidence of Aplastic Anaemia in Thailand. However, complete surveys of incidences of Aplastic Anaemia in pesticide maker or users and farmers are not complete and their cause-effect relationship is not fully explored. Nevertheless, these studies provided data of signs of Aplastic Anaemia but not up to the fatal limit [60-64]. In contrast, pesticide induced Aplastic Anaemia has been recorded to be fatal in countries like India and other Asian countries which may be accounted for due to the inadequate protective measure in the later cases. Prolonged Pesticide use among farmers from villages showed features of Acquired Aplastic Anaemia.

The mechanism through which the pesticides are absorbed through the skin with the resultant induction of Aplastic Anaemia is not fully explored [65-69] but observations indicated that the induction may follow the route of inhalation, absorption through hair follicles, bare footed skin and the body skin surface. Involvement of vasculoendothelial system is more probable but the possibility of direct absorption in the blood or inside the bone marrow can not be ruled out. A systematic study of blood pesticide level and those in other organ tissue system might have more added values in toxicological studies in pesticide victims. The absorption of pesticides in the bone marrow, although remains a question, we have experience to visualize disorganized bone marrow niche structure in the concerned subjects.

A number of such individuals were reported to have clinical feature of Aplastic Anaemia in our outpatient department. They were investigated for complete haemogram, bone marrow status and bone marrow culture profile. The findings suggested typical symptoms of Aplastic Anaemia (Table 1A, 1B, 1C).

Table IA. Peripheral Blood Haemogram of Healthy Volunteers (n=5) (with informed consent)

Patient	Age/Sex/ Occupation	Hb (gm/dl)	PCV	Ret. $10^3/\mu l$	WBC $10^3/\mu l$	ANC $10^3/\mu l$	Platelet $10^5/\mu l$	Diagnosis
1	24/M/S	14.0	44	0.92	7.8	4.3	3.5	Normal
2	25/M/S	14.7	44	0.99	8.4	4.6	3.6	Normal
3	21/M/G	13.9	42	0.58	6.9	3.4	2.1	Normal
4	32/M/G	14.2	41	0.60	7.2	3.8	2.7	Normal
5	24/M/S	15.0	43	0.90	8.8	5.1	3.8	Normal

G = general profession, S = student.

Table IB. Peripheral Blood Haemogram of Patients with Aplastic Anaemia
(Non-Pesticide induced aplastic anaemia, NPAA)(n=10) (With informed consent)

Patient	Age/Sex/ Occupation	Hb (gm/dl)	PCV	Ret. $10^3/\mu l$	WBC $10^3/\mu l$	ANC $10^3/\mu l$	Platelet $10^5/\mu l$	Diagnosis
1	31/F/HW	5.2	10	0.13	1.6	0.26	0.51	NPAA
2	21/M/G	4.9	11	0.17	1.7	0.28	0.53	NPAA
3	18/M/S	6.0	10.8	0.14	1.9	0.26	0.56	NPAA
4	17/F/S	6.4	10.2	0.16	1.8	0.27	0.54	NPAA
5	26/M/G	5.8	10.9	0.17	1.7	0.26	0.55	NPAA
6	20/M/S	6.1	11	0.19	1.8	0.25	0.49	NPAA
7	24/M/G	5.4	10.5	0.14	1.7	0.27	0.49	NPAA
8	29/M/G	5.8	10.8	0.16	1.9	0.25	0.51	NPAA
9	28/M/G	4.9	9.0	0.17	1.8	0.24	0.54	NPAA
10	19/F/S	5.6	11.0	0.18	1.6	0.27	0.53	NPAA

HW = house wife, G = general profession, S = student.

Table IC. Peripheral Blood Haemogram of Patients with Acquired Aplastic Anaemia
(AAA; Pesticide victims)(n=18) (With informed consent)

Patient	Age/Sex/ Occupation	Hb (gm/dl)	PCV	Ret. $10^3/\mu l$	WBC $10^3/\mu l$	ANC $10^3/\mu l$	Platelet $10^5/\mu l$	Diagnosis
1	18/M/Farmer	3.3	9	0.16	1.9	0.23	0.45	AAA
2	22/M/Farmer	4.2	11	0.18	1.7	0.22	0.44	AAA
3	26/M/Farmer	3.9	10	0.19	2.0	0.22	0.5	AAA
4	28/M/Farmer	4.2	11	0.17	1.8	0.23	0.49	AAA
5	16/F/Farmer	3.6	10	0.16	1.6	0.21	0.44	AAA
6	31/M/Farmer	4.3	12	0.16	1.8	0.22	0.46	AAA
7	31/M/Farmer	5.1	12	0.19	1.6	0.21	0.46	AAA
8	40/M/Farmer	4.8	12	0.18	1.8	0.21	0.47	AAA
9	25/F/Farmer	4.6	11	0.19	1.9	0.24	0.49	AAA
10	27/F/Farmer	4.3	11	0.16	1.6	0.23	0.48	AAA
11	28/F/Farmer	3.9	10	0.17	1.6	0.22	0.48	AAA
12	36/M/Farmer	3.8	9	0.18	1.8	0.21	0.46	AAA
13	26/M/Farmer	4.9	12	0.16	1.7	0.21	0.45	AAA
14	32/M/Farmer	5.1	12	0.15	2.0	0.21	0.43	AAA
15	33/M/Farmer	3.7	9	0.14	2.0	0.22	0.43	AAA
16	45/M/Farmer	4.4	11	0.19	1.9	0.24	0.44	AAA
17	40/M/Farmer	3.7	10	0.11	1.6	0.23	0.46	AAA
18	42/M/Farmer	3.8	9	0.12	1.7	0.23	0.47	AAA

The etiology of induction of Aplasia in these patients was found to be a long drawn pesticide use by these individuals in the agricultural field. The investigations furnished that pesticide-spraying technique was without any precautionary measure wherein they inhaled pesticides or received the same by skin absorption. We isolated and analyzed the bone marrow matter as disorganized and confirmed bone marrow aplasia as a result of pesticide toxicity.

Although a number of investigations denied the destructive role of pesticides on the microenvironmental factors, our observations were in agreement that stem - stromal imbalance might be an important cause of bone marrow aplasia under the event. *In-vitro* studies with microenvironmental supplementation revealed stem - stromal functional recovery [28].

THE CURRENT HYPOTHESIS

Much has been told about the bone marrow microenvironment, which has been considered to have a significant physiological role to maintain the cellular balance in the periphery [37, 39, 40, 41, 70, 71]. The present concept of bone marrow "niche" [64] relates a more scientific concept in favour of cellular generation from pluripotent stem cells. The "niche theory" elaborated the intricate relationship in between the stem and stromal cells [72], their interdependence concerned over the network function with the cytokines, growth factors and the sequential maturation of the pluripotent cells towards healthy functional cells [41, 73]. Other observations provided the evidence for the involvement of matrix components in haemopoietic stem cell (HSC) regulation and further emphasize the importance of regulating anchorage and quiescence as essential features of "niche" function [74].

Investigations conducted in our laboratory both in experimentally induced 'Aplastic Anaemia' mice and in pesticide victims (patients) amply confirm the influence of microenvironmental association factor in the developmental and functional kinetics of stem cells in culture [28]. The observations showed that the association factors, which are the integral components of the niche as well, might influence the above functions. The cord blood derived plasma factors (CBPF) are the representative nourishing molecules for the stem cell development and function, as authenticated. These discussions provide ample evidences towards the fact that stem stromal interdependence is the basis of the spontaneous "self renewal system" of the stem cells and that the functional maturation are directed by the stromal secretions or by direct involvement for stem cell generation. In our laboratory we have been able to show that stromal involvement in terms of supporting base for stem cell generation or maintenance of stemness [75] and the stromal stimulation for stem cell generation in *in-vitro* culture are the pre- requisites of healthy bone marrow functions (Figure 3a, 3b, 3c).

A number of electron microscopic studies of *in-vitro* culture of bone marrow cells revealed the stromal cells as the parent supportive cells to the stem cell progeny representing the "niche" system (Figure 1b, 1c). Furthermore, evidences from the bone marrow of mice with experimentally induced Aplastic Anaemia showed series of stromal derangement (Figure 2b, 2c) together with distorted "niche" structure leading to deficient or no generation of stem cells. This might have an important bearing towards the initiation of bone marrow degenerative processes, wherein supplementation of stromal marrow microenvironment (stromal plus) can be a useful component for supportive therapy [29]. These need extensive and meticulous studies to confirm this speculation and reveal the actual scenario within the bone marrow.

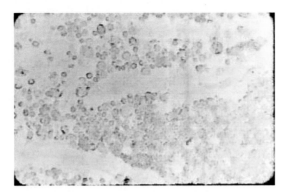

Figure 3a. Ten days culture of pluripotent stem cells, isolated from the bone marrow of normal mice: shows microenvironmental coexistence of stromal cells, stromal precursors and pluripotent stem cells.

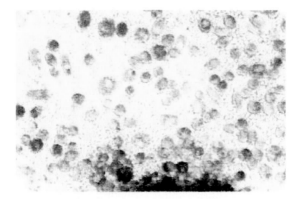

Figure 3b. Explant culture of stroma directed stem cell generation: interdependence of stem and stroma and stromal precursors is revealed on day ten.

Figure 3c. The stem stromal culture of bone marrow from Aplastic Anaemia induced mice: the inappropriate and deceased cellular kinetics is evidenced.

CONCLUSION

The intriguing fact that resounds again and again is that the condition of stem stromal balance is essential in the formation and the maintenance of a "healthy niche" which *inter alia* serves as the complete machinery for haematopoietic generation, wherein different cytokines and growth factors serve as the fuel of the machinery. Thus we emphasize that supplementation of only fuel i.e. cytokines and growth factors may not activate the intricate damages in the machinery as we find in experimental aplasia; here repair of the machinery is the first requirement which may be possible with nourishment, structural balance of associated components etc. Our experience in case of Acquired Aplastic Anaemia (the pesticide victims) suggested that Cord blood derived Plasma Factor (CBPF) provided the appropriate nourishment to repair the "Niche Machinery" which could thereafter utilize the cytokines and growth factors as the fuel for healthy differentiation and proliferation of stem cells.

Thus healthy stem cell generation can be described as a chain of events as: "stem-stromal balance→ the niche network →nourishment→cytokines / growth factors→regulated BM→healthy stem cell generation".

ACKNOWLEDGEMENT

The authors wish to acknowledge Department of Environment, Department of Science and Technology, Government of West Bengal for the kind sponsorship. Thankful acknowledgements are also due to the Director, Calcutta School of Tropical Medicine for the facilities provided. Grateful acknowledgements are also due to the members of the Institutional Ethical Committee who kindly provided the approval for using human and animal samples.

REFERENCES

[1] Brodsky, RA. Biology and management of acquired severe aplastic anaemia. *Current opinion in Oncology*, 1998, 10, 95-99.
[2] Stohlman, F. Jr. Aplastic Anaemia. *Blood*, 1972, 4, 282.
[3] Ehrlich, P. Uber einen Fall von Anamie mit Bemerkungen uber regenerative Veranderungen des Knochenmarks. Charite-Annalen. 1888, 13, 300.
[4] Chauffard, M. Un cas d'anaemie pernicieuse aplastique. *Bull. Mem. Soc. Med. Hop.* Paris, 1904, 21, 313.
[5] Juneja, HS; Lee, S; Gardner, FH. Human long term bone marrow cultures in aplastic anaemia. *Int. J. Cell Cloning*, 1989, 7, 129-135.
[6] Testa, N; Headry, J; Molineux, G. Long term bone marrow damage after cytotoxic treatment: Stem cells and microenvironment. In: Testa NG, Gale RP, eds *Haematopoiesis*, 8th ed, vol. 75. New York, Marcel Dekker Inc., 1988.
[7] Law, Sujata; Maiti, D.; Palit, Aparna; Majumder, D; Basu, K; Chaudhuri, Swapna and Chaudhuri, S. Facilitation of functional compartmentalization of bone marrow cells in leukaemic mice by Biological Response Modifiers: An immunotherapeutic approach; *Immunology Letters*, 2001, 76 (3), 145-152.

[8] Law, Sujata ; Maiti, D.; Palit, A. and S. Chaudhuri. Role of Biomodulators and involvement of Protein Tyrosine Kinase on stem cell migration in normal and leukaemic mice. *Immunology letters*, 2003, 86(3), 287-290.

[9] Chaudhuri, S and Law, Sujata. Stem cells and frontiers of therapeutic advances in cancer. *J. Exp. Clin. Can. Res.*, 2005, 24(2), 207-215. (Review article)

[10] Law, Sujata; Begum, B; Chaudhuri, S. Pluripotent bone marrow cells in leukaemic mice elicit enhanced immune reactivity following Sheep erythrocyte administration in-vivo. *J. Exp. Clin. Cancer Res.*, 22 (2), 421-429, 2003.

[11] Maciejewski, JP; Selleri, C; Sato, T, et al. A severe and consistent deficit in marrow and circulating primitive haematopoietic cells (long-term culture initiating cells) in acquired aplastic anaemia. *Blood*, 1996, 88, 1983-1991.

[12] Scopes, J; Bagnara, M; Gordon-Smith EC, et al. Haemopoietic progenitor cells are reduced in aplastic anaemia. *Br. J. Haematol.*, 1994, 86, 427-430.

[13] Manz, CY; Nissen, C; Wodnar-Filipowicz, A. Deficiency of $CD34^+$ c-kit$^+$ and $CD34^+CD38^-$ haematopoietic precursors in aplastic anaemia after immunosuppression treatment. *Am. J. Haematol.*, 1996, 52, 264-274.

[14] Yoshida, K; Miura, I; Takahashi, T, et al. Quantitative and Qualitative analysis of stem cells of patients with aplastic anaemia. *Scand. J. Haematol.*, 1983, 30, 317-323.

[15] Maciejewski, JP; Kim, S; Sloand, E, et al. Sustained long-term haematologic recovery despite a marked quantitative defect in the stem cell compartment of patients with aplastic anaemia after immunosuppressive therapy. *Am. J. Haematol.*, 2000, 65, 123-131.

[16] Knospe, WH; Crosby ,WH. Aplastic anemia: a disorder of the bone marrow sinusoidal microcirculation rather than stem-cell failure? *Lancet*, 1971, 1, 20-22.

[17] Gluckman, E; Rokicka-Milewska, R; Hann, I, et al. Results and follow-up of a phase III randomized study of recombinant human-granulocyte stimulating factor as support for immunosuppressive therapy in patients with severe aplastic anaemia. *Brit. J. Haematol.*, 2002, 119, 1075-1082.

[18] Maciejewski, JP; Anderson, S; Katevas, P, et al. Phenotypic and functional analysis of bone marrow progenitor cell compartment in bone marrow failure. *Brit. J. Haematol.*, 1994, 87, 227-234.

[19] Koijima, S. Haematopoietic growth factors and marrow stroma in aplastic anaemia. *Int. J. Haematol.*, 198, 68, 19-28.

[20] Marsh, JC. Haematopoietic growth factors in the Pathogenesis and for the treatment of aplastic anaemia. *Semin. Haematol.*, 2000, 37, 81-90.

[21] Koijima, S; Matsuyama, T. Stimulation of granulopoiesis by high-dose recombinant human granulocyte colony-stimulating factor in children with aplastic anaemia and very severe neutropenia. *Blood*, 1994, 83, 1474-1478.

[22] Sonada, Y; Ohno, Y; Fujii, H, et al. Multilineage response in aplastic anaemia patients following long term administration of filgrastim (recombinant human granulocyte colony stimulating factor). *Stem Cells*, 193, 11, 543-554.

[23] Ganser; A; Lindemann, A; Seipelt, G, et al. Effects of recombinant human interleukin-3 in aplastic anaemia. *Blood*, 1990, 76, 1287-1292.

[24] Nimber, SD; Paquette, RL; Ireland, P., et al. A phase I/II study of interleukin-3 in patients with aplastic anaemia and myelodysplasia. *Exp. Haematol.*, 1994, 22, 875-880.

[25] Bargetzi, MJ; Gluckman, E; Tichelli, A, et al. Recombinant human interleukin-3 in refractory severe aplastic anaemia: a phase I/II trial. *Brit. J Haematol.*, 1995, 1, 306-312.

[26] Kurzrock, R; Paquette, R; Gratwohl, A et al. Use of stem cell factor (Stemgen, SCF) and filgrastim (G-CSF) in aplastic anaemia patients who have failed ATG/ALG therapy. *Blood*, 1997, 90 [Suppl-1], 173.

[27] Gordon, MY; Greaves, MF. Physiological mechanisms of stem cell regulation in bone marrow transplantation and haemopoiesis. *Bone Marrow Transplant*, 1989, 4, 335-338.

[28] Law, Sujata; Basu K; Banerjee, S; Begum, B and Chaudhuri, S. Cord blood derived plasma factor (CBPF) potentiates the low Cytokinetic and Immunokinetic profile of Bone Marrow Cells in pesticide victims suffering from Acquired Aplastic Anaemia. *Immunological Investigations*, 2006, 35, 209-225.

[29] Basu, K. Structure function correlationship of the pluripotent cells of Bone Marrow in health and disease: An immunological approach towards the haematopoietic microenvironment versus biological response modifiers. Ph. D. Thesis. 2003, Jadavpur University, Kolkata, India.

[30] Ershler, WB; Ross, J; Finlay, JL, et al. Bone marrow microenvironment defect in congenital hypoplastic anaemia. *N. Eng. J. Med.*, 1980, 302, 1321-1327.

[31] Tavassoli, M and Friedenstein, A. Haemopoietic stromal microenvironment. *Am. J. Haemat.*, 1983, 15, 195-203.

[32] Dexter, TM; Moore, MA. In vitro duplication and "cure" of haemopoietic defects in genetically anaemic mice. *Nature* 197, 269, 412-414.

[33] Dexter, TM. Stromal cell associated haemopoiesis. *J. Cell Physiol.*, 1982, Suppl-1, 87-94.

[34] Dexter, TM; Allen, TD and Lajtha, LG. Conditions controlling the proliferation of haemopoietic stem cells in vitro. *J. Cell Physiol.*, 1977, 91, 335-344.

[35] Huss, R . Isolation of primary and immortalized CD34⁻ Haematopoietic and mesenchymal stem cells from various sources. *Stem Cell* 2000, 18, 1-9.

[36] Dexter, TM; Spooncer, E; Verga, J. et al. Stromal Cells and diffusible factors in the regulation of haemopoietic cells development. In: Killman SA, Cronkite EP, Muller-Berat CN, eds. *Haemopoietic stem cells*. Copenhagen: Munksgaard, 1983, 303-318.

[37] Schofield, R. The relationship between the spleen colony-forming cell and the haematopoietic stem cell. *Blood Cells*, 1978, 4, 7-25.

[38] Spradling, A; Drummond-Barbosa, D; Kai,T. Stem cells find their niche. *Nature*, 2001, 414, 98-104.

[39] Singer, JW; Slack, JL; Lilly, MB and Andrews, DF. Marrow stromal cells: response to cytokines and control of gene expression. In: Long MW and Wicha MS eds. *The Haematopoietic Microenvironment*. Baltimore and London Johns Hopkins University Press, 1993, 127-151.

[40] Moore, KA and Lemischka, IR. Stem cells and their niches. *Science*, 2006, 311, 1880-1885.

[41] Scadden, DT. The Stem Cell niche as an entity of action. *Nature*, 441, 1075-1079, 2006.

[42] Hurlay, RW; McCarthy, JB; Verfaillie, CM. Direct adhesion to bone marrow stroma via fibronectin receptors inhibits haematopoietic progenitor proliferation. *J. Clin. Invest.* 1995, 96, 511- 519.

[43] Tavassoli, M; Maniatis, A; Crosby, WH. Induction of sustained haemopoiesis in fatty marrow. *Blood,* 1974, 43, 33-38.

[44] Islam, A. Do bone marrow fat cells or their precursors have a pathogenic role in idiopathic aplastic anaemia? *Med. Hypotheses*, 1988, 25, 209.

[45] Harrison, DE. Use of genetic anaemias in mice as tools for haematological research. *Clin. Haematol.*, 1979, 8, 239-262.

[46] Russell, E. Hereditary anaemias of the mouse: a review for geneticists. *Adv. Genet.*, 1979, 200, 357.

[47] Chen, J; Brandt, JS; Ellison, FM; Calado, RT; Young, NS. Defective stromal cell function in a mouse model of infusion-induced bone marrow failure. *Exp. Haematol.*, 2005, 33(8),901-908.

[48] Young, NS; Issaragrasil, S; Chich, CW, et al. Aplastic anaemia in the Orient. Br J Haematol, 1986, 62, 1-6.

[49] Young, N. Acquired aplastic anaemia. In: Young, NS ed. *Bone marrow failure syndromes*. Philadelphia: WB Saunders, 2000, 1-46.

[50] Young, N. Drugs and chemicals as agents of bone marrow failure. In: Testa N, Gale RC, eds. *Haematopoiesis; long term effects of chemotherapy and radiation*. Vol 131, New York: Marcel Dexter Inc. 1988.

[51] Gordon-Smith, EC. Aplastic anaemia- etiology and clinical features. *Baillieres Clin Haematol.* 1989, 2, 1-18.

[52] Young, N. Drugs and Chemicals. In: Young NS, Alter B, eds. *Aplastic anaemia acquired and inherited*. Philadelphia: WB Saunders, 1994, 100-132.

[53] Issaragrisil, S; Kaufman, DW; Anderson, T. et al. The epidemiology of aplastic anaemia in Thailand. *Blood*, 2006,107, 1299-1307.

[54] Young, NS. Acquired aplastic anaemia. *Ann. Intern. Med.*, 2002, 136, 534-546.

[55] Muir, KR; Chilvers, CE; Harriss, C., et al. The role of occupational and environmental exposures in the etiology of acquired severe aplastic anaemia: a case control investigation. *Br. J. Haematol.*, 2003, 123, 906-914.

[56] Scott, J; Cartwright, G; Wintrobe, M. Acquired Aplastic anaemia: an analysis of thirty-nine cases and review of the pertinent literature. *Medicine*, 1958, 36, 119.

[57] Aoki, K; Ohtani, M; Shimizu, H. Epidemiological approach to the etiology of Aplastic anaemia. In: Hibiono S, Tataku F, Shahidi NT, eds. *Aplastic anaemia*. Baltimore University Park Press 1978.

[58] Williams, DM; Lynch, RE; Cartwright, GE. Drug induced aplastic anaemia. *Semin Haematol*, 1973, 10, 195-223.

[59] Whang K. Aplastic anaemia in Korea: a clinical study of 309 cases. In: Hibino S, Takaku F, Shahidi NT eds. *Aplastic anaemia*. Baltimore University Park Press, 1978, 225.

[60] Stormont, R. Pharmacologic and toxicologic aspects of DDT (Chlorophenothane U.S.P.). *JAMA*, 1951, 135, 728.

[61] Woodliff, HJ; Connor, PM; Scopa, J. Aplastic anaemia associated with insecticides. *Med. J. Aust.*, 1966, 1, 628-629.

[62] Friberg, L; Martensson, J. Case of Panmyelophthisis after exposure to chlorophenothane and benzene hydrochloride. *AMA Arch. Ind. Hyg.*, 1953, 8, 166.

[63] Wang, HH; Grufferman, S. Aplastic anaemia and occupational pesticide exposure: a case control study. *J. Occup. Med*, 1981, 23, 364-366.

[64] Linos, A; Kyle, RA; O'Fallon, WM, et al. A case control study of occupational exposures and leukemia. *Int. J. Epidemiol.*, 1980, 9, 131-135.

[65] Sciences International. Toxicological Profile for Stoddard Solvent: U.S. Department of Health and Human Services, Public Health Service, Agency for Toxic Substances and Disease Registry; 1995.

[66] Fleming, LE ; Timmeny, W. Aplastic anaemia and pesticides: an etiologic association ? *J. Occup. Med.*, 1993, 35, 1106-1116.

[67] Sanchez-Medal, L ; Castanedo, JP ; Garcia-Rojas, F. Insecticides and aplastic anaemia. *N. Engl. J. Med.*, 1963, 269, 1365-1367.

[68] Young, NS ; Maciejewski, J . The pathophysiology of acquired aplastic anaemia. *N. Engl. J. Med.*, 1997, 336, 1365-1372.

[69] Aksoy, M ; Erdem, S ; Dincol, G ; Bakioglu, I; Kutlar, A. Aplastic anaemia due to chemicals and drugs: a study of 108 patients. *Sex Transm. Dis.*, 1984, 11, 347-350.

[70] Weiss, L. The Haematopoietic microenvironment of bone marrow: an ultrastructural study of the stroma in rats. *Anat. Rec.*, 1976, 186, 161-84.

[71] Weiss, L and Sakai, H. The haematopoietic stroma. Am *J Anat.*, 1984, 170, 447-463.

[72] Li,L; Xie, T. Stem cell niche: Structure and function. *Annu. Rev. Cell Dev. Biol.*, 2005, 21, 605-631.

[73] Zhang, J; Niu, C; Ye, L et al. Identification of the haematopoietic stem cell niche and control of the niche size. *Nature*, 425, 836-841, 2003.

[74] Heissig, B; Hattori, K; Dias, S et al. Recruitment of Stem and Progenitor cells from the bone marrow niche requires MMP-9 mediated release of kit-ligand. *Cell*, 2002, 109, 625-637.

[75] Constantencscu, S. Stemness, Fusion and renewal of haematopoietic and embryonic stem cells. *J. Cell. Mol. Med.*, 2003, 7, 103-112.

In: Developments in Stem Cell Research
Editor: Prasad S. Koka

ISBN: 978-1-60456-341-2
© 2008 Nova Science Publishers, Inc.

Chapter 10

NATURAL KILLER CELL RECEPTOR NKG2A/HLA-E INTERACTION DEPENDENT DIFFERENTIAL THYMOPOIESIS OF HEMATOPOIETIC PROGENITOR CELLS INFLUENCES THE OUTCOME OF HIV INFECTION

Edmond J. Yunis[*][a], *Viviana Romero*[a], *Felipe Diaz-Giffero*[a],
Joaquin Zuñiga[a,b] *and Prasad Koka*[b]

[a]Department of Cancer Immunology and AIDS, Dana-Farber Cancer Institute and
Department of Pathology, Harvard Medical School, Boston MA
[b] Laboratory of Immunobiology and Genetics, Instituto Nacional de Enfermedades
Respiratorias, Mexico D.F. , Mexico.
[c]Laboratory of Stem Cell Biology, Torrey Pines Institute for Molecular Studies
San Diego, CA 92121, USA

ABSTRACT

HIV infection and its outcome is complex because there is great heterogeneity not only in clinical presentation, incomplete clinical information of markers of immunodeficiency and in measurements of viral loads. Also, there many gene variants that control not only viral replication but immune responses to the virus; it has been difficult to study the role of the many AIDS restricting genes (ARGs) because their influence vary depending on the ethnicity of the populations studies and because the cost to follow infected individuals for many years. Nevertheless, at least genes of the major

[*] **Correspondence:** Edmond J. Yunis, M.D., Department of Cancer Immunology and AIDS, Dana Farber Cancer Institute, Harvard, Medical School, 44 Binney Street, Boston, MA. 02115. Phone: (617) 632 3347 Fax: (617) 632 5151. Email: edmond_yunis@dfci.harvard.edu

histocompatibility locus (MHC) such as HLA alleles have been informative to classify infected individuals following HIV infection; progression to AIDS and long-term-non-progressors (LTNP). For example, progressors could be defined as up to 5 years, up to 11 years or as we describe in this report up to 15 years from infection, and LTNP could be individuals with normal CD4+ T cell counts for more than 15 years with or without high viral loads. In this review, we emphasized that in the studies of ARGs the HLA alleles are important in LTNP; HLA-B alleles influencing the advantage to pathogens to produce immune defense mediated by CD8+ T cells (cognate immunity). Our main point we make in this report is that contrary to recent reports claiming that this dominant effect was unlikely due to differences in NK activation through ligands such as HLA-Bw4 motif, we believe that cognate immunity as well as innate inmmunity conferred by NK cells are involved. The main problem is that HLA-Bw4 alleles can be classified according the aminoacid in position 80. Isoleucine determines LTNP, which is a ligand for 3DS1. Such alleles did not include HLA-B*44, B*13 and B*27 which have threonine at that position. The authors have not considered the fact that in addition to the NK immunoglobulin receptors, NK receptors can be of the lectin like such as NKG2A/HLA-E to influence the HIV infection outcome. HLA-Bw4 as well as HLA-Bw6 alleles can be classified into those with threonine or methionine in the second position of their leader peptides. These leader peptides are ligands for NKG2A in which methionine influences the inhibitory role of NKG2A for killing infected targets. Functional studies have not been done as well as studies of these receptors in infected individuals. However, analyses of the leader peptides of HLA-B alleles in published reports, suggested that threonine in the second position can explain the importance of HLA-B*57, B*13, B*44 as well as certain Bw6 alleles in LNTP. In addition, we analyzed the San Francisco database that was reported and found that the association of HLA-B alleles with LNTP or with progressors can be due to the presence of threonine or methionine in their second position. Therefore, studies of outcome of HIV infection should include not only mechanisms of cognate immunity mediated by peptides and CD8+ T cells but also, NK receptors of two types, NKG2A as well as 3DS1. We propose that the SCID mouse should be used to understand mechanisms mediated by many of the ARGs especially the importance of thymus derived cells as well as NK receptor interactions with their ligands in this experimental animal transplanted with human stem cells, thymus or NK cells obtained from individuals of known HLA genotypes.

I. INTRODUCTION

AIDS is not generally considered a genetic disease because there is a great heterogeneity of clinical presentation in part determined by gene variants that control to same extent virus replication and immunity [1-3]. Noteworthy of mentioning is the fact that there is not much information about genes that protect individuals exposed to the virus and do not get infected. Also, it is important to mention that those that get infected develop pathology at different times following the infection, for example those who died from AIDS at 1-5 years from infection, but the definition of such fast progressors should include those where the immunodeficiency and progression to AIDS could go up to 10-11 years or more from the time of the infection. More difficult perhaps, is to define long-term infection since there are patients that progress to immunodeficiency 15-20 or more years after infection and a group of the long-term non progressors that maintain normal CD4+ T cell counts and absence of viremia for more than 20 years without treatment. The heterogeneity of clinical presentation cannot be completely explained, but it is possibly related to the large number of genetic and non-genetic contributing factors: innate, humoral and cell mediated immune responses [4] and the AIDS restricting genes (ARGs). In addition, longitudinal studies have an additional difficulty because some patients are treated with antiretroviral drugs to prevent or diminish the possible progression to AIDS.

Our review will attempt to emphasize that studies of CD8+ T cell immunity should be performed in conjunction with NK receptor/ligand interactions and their variants. We will summarize evidence that HIV infection involves several immune mechanisms resulting in long term non progression or progression to AIDS in which intrinsic, innate and cognate immunity are important. However, genetic associations with protection from or progression to disease are complex and could vary, related to the ethnicity of the populations studied [5].

II. IMMUNE FUNCTIONS IN HIV INFECTION

A. Cognate Immunity

Cognate immunity has been extensively studied involving primarily CD4+ and CD8+ T cells interacting with MHC alleles presenting HIV-1 peptides on antigen presenting cells [6-12]. Such cells may mediate protection by production of cytotoxic cells or cytokines such as IFN gamma and IL-12 [13]. Macrophages [14], DC and plasmocytoid DCs are type 1 IFN secretion inducers of IL-12. However, several viruses escape MHC class I restricted cytotoxic T lymphocytes (CTL) responses by down-regulating their expression on infected cell surface [15]. In this regard, protection from infection may have genetic basis but it varies in different populations [5]; these associations are complex and difficult to assess because in most cases is difficult to know the date of incidence of infection, the date of presence of viremia, or the date of decrease of CD4 counts. In many cases the studies could be difficult to reproduce because of genetic stratification and population size of case-control studies. MHC studies, in many cases, are incomplete and without strict criteria for population sizes and epidemiological parameters that explain the lack of consistencies of the alleles associated with disease progression [16]. Also, some of the MHC associations maybe present because of genes that are in linkage disequilibrium with HLA alleles, for example TNF [17], that are involved in the regulation of level of HIV-1 viremia since it causes increased viral replication in infected monocytes. This occurs through the activation of the transcription factor NFkB, which binds to the HIV-1 long terminal repeat, causing increased levels of HIV-1 transcription [18-20].

HLA Alleles, and Clinical Outcome of HIV-1 Infection

HLA alleles of class I (A, B and C) and II (DR, DQ and DP) have a large variation between individuals and populations, which could provide recognition of virus agents to which they have been exposed [4]. Since HIV infects immune cells to produce proliferation, spread and CD4+ T lymphocyte damage, the HLA alleles could influence the time from infection to AIDS progression [21]. For example, HLA-Cw4, a ligand for KIR 2DL1 influences the time to develop AIDS, [22]. These findings could have been due to genes in random association with HLA-B and C alleles such as TNF [17]. The contribution of HLA-B*35 alleles has also been postulated [23]. More importantly, HLA influence was observed in relation to HIV-1 subtypes where the HLA recognition motifs are associated to AIDS survival. Two HLA alleles, B*27 and B*57, have been reproducibly associated with long-term non-progression of AIDS [3, 24]. A controversial subject is the role of zygosity of HLA-A alleles in the progression to AIDS. For example HLA class I homozygotes for several alleles progress to AIDS faster than those heterozygous [23, 24]. Further, groups of alleles or supertypes grouped according the B pocket showed association with clinical outcome [16, 25]. The role of zygosity for HLA-Bw6 and HLA-Bw4 has been controversial, Bw6 was found associated with AIDS in Caucasians [25, 26] and Chinese [27] while the Bw4/Bw4 was associated with LTNP in Caucasians but not in Africans [28]. Therefore, genetic markers, ARGs result from different evolutionary effects that would explain that ethnicity and protection from AIDS and genetic effects on clinical outcome vary among populations [4, 5].

B. Innate Immunity

This type of immunity is mediated by a newly discovered endogenously expressed proteins that provide defenses against retroviral infection such as HIV-1 and murine leukemia virus (MLV) [29-31]. These proteins probably prevent the entry of HIV-1 in to cells by interaction with CD4, Natural Killer (NK) cells, monocytes and dendritic cells. In addition, several molecules secreted by these cells can protect against such infection, for example IL-12, chemokine receptors CCR4 and CCR5. In this regard, most HIV strains use the CCR5 as a co-receptor and thereby are sensitive to inhibition by the ligands of this receptor; also deletion of CCR5 prevents viral entry into cells [2]. HIV-1 infection is potently blocked in rhesus macaques by the cytoplamic component protein tripartite motif 5 (TRIM5). Interestingly, the TRIM5 protein in rhesus macaque is 87% identical to its human homolog. However, human TRIM5 proteins do not restrict HIV-1 [31]. Remarkably, one amino acid change in the protein sequence of human TRIM5 leads to full activity against HIV-1 infection [32, 33]. Even though, several SNPs for human TRIM5 have been identified, none of them corresponded to this specific residue. Differently, the endogenously expressed human cytidine deaminase APOBEC3G potently blocks HIV-1 infection [32]. However, the HIV-1 accessory protein vif is able to overcome this restriction allowing productive infection. Finally, the endogenously expressed mouse protein Fv1 potently block infection of MLV [29]. Overall, this demonstrates that during evolution, humans developed series of specie-specific blocks for retroviral infection.

In an extensive genetic epidemiological study, SNPs of TRIM5 were identified. In African Americans four alleles exhibited different frequencies in HIV-1 infected and uninfected individuals. SNP2 in the non-coding exon 1 and SNP3 in intron 1 were associated with increased risk of infection. [34]. These finding suggested that any modification of infection susceptibility afforded by particular TRIM5 alleles may be restricted to particular populations or types of exposure. There are population specific effects for AIDS modifying genes [33-36]. For example, the role of TRIM5 polymorphism in Europeans, the TRIM haplotype containing 136Q exhibited increased frequency among HIV-1 infected subjects compared with seronegative exposed individuals [34], which contrasts with the opposite results reported in African Americans [35, 36]. These findings suggest that either the genetic background, non-random association with other genes or the type of exposure may influence susceptibility to HIV-1 infection. However, TRIM5 variants do not influence disease progression. As it will be pointed out below, there are many genes that limit AIDS and such influence may be present in some ethnicities but not in all. This problem stresses the fact that research is needed to determine their frequencies in several ethnicities in order to include sufficient individuals in the calculations of statistical power.

According to estimations there are more than 90% of the genetic and non-genetic influences on AIDS progression that are still undiscovered. Less known is the estimation of these factors influencing genetic susceptibility to be infected by HIV-1. It was calculated that 21.1% of individuals infected who did not develop AIDS for 11 more years did so because they carried one of more protective ARGs. Among those that develop AIDS rapidly, within 5.5 years of HIV-1 infection, the single ARG effects were modest (2.4-8.1%) but the cumulative ARGs associated with rapid progression is high (40.9%). However, the estimates are approximately 10% for both groups including long-term non-progressors and progressors to AIDS. Available databases, although informative, possess problems for analyses, because during the patients follow up some of them are recruited after they have been treated with anti-retroviral drugs during the first 10-11 years from the time of infection. Thus the definition of immunodeficiency is based on CD4+ T cell counts only without consideration of viral loads. Also some patients without treatment maintain normal CD4+ T cell counts beyond 15 years from infection while they may have high viral loads.

NK cells

Studies of HIV-1 uninfected persons showed enhanced activity of NK functions despite many years of high-risk exposure demonstrating the importance of NK cells in immunity against HIV-1 infection [37]. NK cells functions involve receptors that interact with HLA ligands. HLA alleles are therefore important in disease progression because target infected cells are more susceptible to NK killing. For example, *nef*, a product of HIV-1 is known to diminish the levels of HLA-A and HLA-B expression of infected cells [38], whereas HLA-C, which is one ligand for NK receptors, is poorly expressed naturally [39]. By contrast, *nef* increases the level of HLA-E on the surface of infected cells [40]. Also, HLA-A expression is higher than HLA-B [39] and HLA-B expression is more inducible by IFN alpha than the HLA-A [41]. Therefore, in the environment of the HIV infection there would be more contribution of HLA-B alleles or leader peptides from HLA-B as well as that of HLA-E with HLA-B leader peptides to function in their interaction with NK receptors.

There are two different kinds of NK receptors, immunoglobulin-like such as the inhibitory and activating receptors that include the KIRs and also lectin-like such as the NKG2 receptors. The NKG2A is inhibitory and has as a ligand HLA-E and the NKG2D is an activating-type receptor and has as a ligand the MICA alleles for function, killing of infected cells or cancer cells [42, 43]. Also, some KIRs are expressed on a subset of NK cells with a memory phenotype [44] that suggests that they may regulate T cell as well as NK cell activity. Masking of inhibitory NK receptors on CTLs from HIV infected individuals by monoclonal antibody has produced increases of HIV-1 specific CTL activity [12] suggesting possible involvement of NK cells.

As proposed in this report, HLA-E and its interaction with NKG2A needs to be investigated in both high risk persons exposed to HIV that have or do not have infection or their role in the clinical outcome following HIV-1 infection. However, the MICA alleles, which are ligands for the lectin activating receptor NKG2D, were not associated with HIV-1 clinical progression [17].

It is well known that HLA-E folds to function, depending on the leader peptides from HLA-class I alleles (A, C and B). The critical aminoacid is methionine in second position, whereas threonine is accompanied by poor folding and defective function [45]. The leader peptides with methionine in the second position (VMAPKTVLL and VMAPRTLL) induce higher levels of HLA-E expression than those with threonine at Position 2. Also, they exhibited high affinity for soluble CD4 NKG2A molecules [45-47]. Despite the high expression of HLA-E due to nef, the presence of threonine in the leader peptides of HLA-B alleles render HLA-E to be poorly expressed [40].

HLA-B Ligands for NK Receptors: HLA-Bw4 and HLA-Bw6 Supertypes and the Leader Peptide of HLA-B alleles as Ligands for the NKG2A Receptor

The HLA-B alleles are the most important in HIV viral progression because they restrict infection via CD4/CD8 [48] and they are ligands for NK receptors, KIRs (3DL1 and 3DS1) [26] and lectin receptor (NKG2A) as we propose in this article.

HLA-Bw4 supertype comprising approximately 40% of HLA-B alleles is the ligand for NK receptors encoded by the Killer immunoglobulin receptor (KIR) gene complex on chromosome 19 [49]. This interaction could be due to loss of inhibition of the inhibitory receptor, 3DL1 in effector cells that causes the function of the activating receptor 3DS1. This was found experimentally, that the combination of homozygosity of HLA-Bw4 and KIR3DS1 epistasis influence AIDS progression [23]. Of interest, this interaction did not include the role of B*27 and B*44 in LNTP. The authors suggested that those alleles are involved in a different mechanism of protection from progression. In this report we are proposing a mechanism that needs to be investigated.

All HLA-A and HLA-C alleles encode HLA-E binding peptides with methionine at second position [50]. Importantly, HLA-C alleles are poorly expressed on the surface of cells [39] and HLA-B has higher basal level of expression than HLA-A, and by far more inducible by IFN gamma and alpha than HLA-A gene [41]. Furthermore, HLA-B alleles are the dominant influence to mediate a possible co-evolution of HIV and HLA [44], that we believe should also include HLA-E loaded with class I peptides.

These findings are based on the fact that HLA-B alleles can be classified according to the presence of thr or met at P2 of the leader peptide [41, 43]. Almost all HLA-Bw4 alleles with the exception of HLA-B*38 encode leader peptides with Thr at P2 and HLA-Bw6 are divided into two groups, those encoding Met at P2 and those encoding Thr at P2 [41, 43, 51] (Table 2).

Statistical Analysis of Influence of Threonine in Second Position in Long Term Non Progressors (LTNP)

Re-analyses of the data published before (Flores et al.) (Table 3) the number of individuals with two copies of Threonine in the group of controllers was significantly higher than in the progressors. 14/20. (0.70) versus 3/19 (0.15) of the non-controllers p: 0.0006 OR: 12.44. Of interest, the frequency of individuals with TT uninfected was 42/108 (0.39). p: 0.010 OR:3.67.

Table 1. The HLA Bw4 and Bw6 motifs and alleles

HLA Supertype		Amino-acid position				Corresponding alleles
		80	81	82	83	
Bw4	Ile80	Ile	Ala	Leu	Arg	B*15 (13, 16, 17, 23, 24) B*2702, **B*3801**, B*4901, B*51, B*5201 B*5301, B*5302, B*57, B*58, B*59
	Thr80	Thr	Ala	Leu	Arg	B*13 (01, 02, 04), **B*3802**, B*44 (02, 03, 04, 05, 06, 21)
		Thr	Leu	Leu	Arg	B*2705, B*2709, B*3701, B*4701
Bw6	Asn 80	Asp	Leu	Arg	Gly	B*07 (02, 03, 04, 05, 06, 09, 10, 14) **B*08**, B*14. B15*(01, 03, 04, 09, 15, 18, 22, 37 45, 46, 13, 16, 17, 23, 24)) B*18, B*35, B*39, B*40, B*41, B*42, B*4501, B*4601, B*48, B*50, B*5401, B*55, B*56, **B*67**, B*73, B*7801, B*8101, B*82

In bold are alleles whose leader peptide has methionine in the second position and the rest have threonine in the second position [50].

Table 2. Comparison of Bw4 and Bw6 frequencies based on the presence of methionine or threonine in second position of HLA-B leader peptides

aa 2nd position	Caucasians	African Americans	Asian Americans
Met	0.30	0.20	0.18
Thr	0.70	0.80	0.82
Met/Met	0.09	0.04	0.04
Met/Thr	0.42	0.32	0.27
Thr/Thr	0.49	0.64	0.69
Supertype alleles			
Bw6	0.61	0.74	0.55
Bw4	0.39	0.26	0.45
Supertype genotypes			
Bw6/Bw6	0.37	0.55	0.30
Bw4/Bw4	0.15	0.07	0.20
Bw6/Bw4	0.49	0.38	0.50

Analyses of the San Francisco cohort reported before also demonstrated that Comparison of LTNP with progressors, either 0-10 years or 0-15 years showed that the presence of TT was significaltly higher in the LTNP than in 0-10 years progressors, 17/22 versus 11/44, p=

0.000005, OR=10.20 and between all progressors (0 to 15 years, 17/22 versus 20/76, p=0.00001, OR= 9.52. Likewise, the frequency of TT was higher in LTNP than in uninfected controls 20/76 versus 130/265, p =0.0004 OR= 2.70. These results should be confirmed using Kaplan-Meier plots to assess more accurately the importance of these findings using outcome of HIV viral infection and need to be studied in relation to the expression of NKG2A receptors on NK cells.

Table 3. Frequencies of leader peptides (two copies) of methionine (M) or threonine (T), influence in HIV progression

aa 2nd position	Intermediate Progressors (10-15 years)	Progressors (0-10 years)	LTNP	Non-infected controls
MM	7/32 (21.8)	14/44 (31.8)	1/22 (4.5)	24/265 (9.0)
MT	16/32 (50.0)	19/44 (43.2)	4/22 (18.2)	138/265 (52.0)
TT	9/32 (28.1)	11/44 (25.0)	17/22 (77.3)	130/265 (49.0)

III. STEM CELL MICROENVIRONMENT IN HIV INFECTION

HIV infection of stem cell microenvironments or niches causes hematopoietic inhibition and hence cytopenias [52-57]. Hematopoietic CD34+ progenitor stem cells are reported to be resistant to HIV-1 infection, *in vitro*, or *in vivo* [58, 59]. Those cells that experienced the indirect effects of HIV-1 infection exhibit inhibition of their multilineage hematopoiesis as determined by colony forming activity ex vivo [58, 60-62]. It is reported that the hematopoietic stem cell microenvironment is damaged due to the indirect effects of HIV-1 infection of the thymocytes on the CD34+ progenitor stem cells but in a reversible manner, in the human fetal Thymus/Liver conjoint hematopoietic organ of the transplanted chimeric severe combined immunodeficiency mouse (SCID-hu) model system [60, 62]. It is therefore highly plausible that this implanted human organ in the SCID-hu mouse, which serves as a niche, not only for thymocyte expansion but also supports hematopoiesis, suffers niche dysfunction due to HIV-1 infection. Continued presence of the $CD34^+$ progenitor stem cells in the infected niche seem to suffer due to exacerbation resulting from persistent virus mediated niche disruption via infection of thymocytes and consequent interactions and signaling network of the hubs. In this microenvironment it is possible that several ARGs could produce different outcomes. This is evident from our previous observation that CCR4 and CCR5- tropic HIV-1 produce variable kinetics of inhibition of hematopoiesis [60].

Less understood is the role of different lymphoid organs in regards to the diminution of T cells in the gastrointestinal niche which is involved in HIV infection clinical outcome especially the genetic markers that contribute to AIDS progression. The intestinal mucosal immune system is an important target of HIV-1 infection and contributes to disease progression. In addition distinct gene expression profiles correlate with clinical outcome [63]. In this regard we wonder if the variable time of outcome of AIDS following HIV infection is related to several innate unknown mechanisms operating in the host. Independent of these is the fact that patients with HIV do not get diagnosed immediately after infection and the time of diagnosis could be variable taking sometimes several years. For example, we had access to the San Francisco database [46] and several individuals recruited for follow-up were first evaluated from 0 to 11 years and therefore the definition of long-term non-progression can only be analyzed after years of the date of known infection. In addition, some of such patients were censored after they had begun treatment. Only in those individuals followed from the time of infection it would be possible to demonstrate the role of genetic markers in viremia, and their participation in producing decreased CD4 counts and/or conversion to AIDS.

The transcription factors such as STAT5A are involved in stem cell self-renewal that precedes multilineage differentiation of CD34+ progenitor stem cells [64-69]. The proto-oncogene of myeloproliferative leukemia also known as thrombopoietin (Tpo) receptor proto-oncogene, c-mpl, is known to promote multi-lineage pluripotent stem cell differentiation of the CD34+ progenitor cells [70-73]. Both STAT5 and c-mpl are important target genes for control and enhancement of stem cell self-renewal and multi-lineage differentiation to reduce or prevent cytopenias induced during HIV infection [74]. We wonder if some individuals have highly regulated expression of STAT5A/B and c-mpl genes in progenitor cells and if in such individuals CD4 cells are generated in higher numbers than in those without such a regulated niche.

IV. DISCUSSION

It is necessary to investigate the role of NK cells in the innate immunity against HIV infection. It is now clear that there are at least three kinds of immune mechanisms involved in the control of viral infections including that of HIV-1: A) intrinsic innate immunity mediated by a group of major defenses against infection by retroviruses, Fv1 and TRIM5 inhibitors, proteins that target incoming retroviral capsids and the APOBEC3 class of cytidine deaminases that hypermutate and destabilize retroviral genomes. These are probably involved in the prevention of HIV-1 into cells and constitute the first level of protection from infection by specific cells, such as dendritic cells, including plasmocytoid dendritic cells. Of course many proteins such as cytokines, IL12, chemokines, and chemokine receptors CCR5 and CXCR4 are important. B) Genetic markers of immune effectors that are important in the outcome of HIV-1 infection; Natural killer, NK, receptors of two kinds, Ig-like such as KIRs and lectin-like such as NKG2 are involved. These two kinds of genetic markers have been described to be involved in the outcome of the HIV-I infection towards long term non-progression and AIDS. In this regard, two reports have described the role of HLA-Bw4/Bw4 as a marker for LTNP and related to that a subgroup of Bw4, isoleucine in position 80 of Bw4 interacting with a KIR receptor, 3DS1 was associated with LTNP [25, 26]. In unpublished work we have re-analyzed the data published before involving the importance of Bw4 in LTNP, discovered that methionine, Met/Met, in the second position of the leader peptide of HLA-B alleles is a marker for progression to AIDS [47]. NK receptors may be involved in the control of viremia; CD94/NKG2A inhibitory receptors [75-77] and the NKG2D stimulatory receptors not only present in NK cells but also in human CD8 lymphocytes [39, 78]. Therefore, NK and CTL activating and inhibitory signals can be provided by NKG2 receptor interaction with MICA and by CD94/NKG2A interaction with HLA-E molecules with the appropriate assembly of peptides [44]. However, the role of NKG2D/MICA in the progression to AIDS was not found [17]. The CD94/NKG2A-HLA-E pathway has not been studied related to HIV-1 infection outcome. It is important to mention that HLA leader peptide sequences with Met at P2 induce significant levels of HLA-E expression compared to those with Thr at P2, the latter failed to confer protection from NK lysis and the HLA-E/peptide complexes exhibited high affinity for soluble CD94/NKG2 molecules [45, 47]. Also, although HLA-A and C alleles encode HLA-E binding peptides with Met at P2, HLA-C alleles are poorly expressed on the surface of cells [39] and *nef* of HIV-1 down-regulates the cell surface expression of HLA-B and A but not C or HLA-E [40]. Furthermore, HLA-A alleles are rarely used to restrict CTL epitopes [48, 79]. In sum, in the environment of HIV-1 infected cells the HLA-leader peptides have more contribution to generate HLA-E binding peptides. This is supported by the reduced HLA-E poor expression by cells transfected with HLA-B*51 or B*58 [80, 81]. Therefore, it is possible that NK receptors are involved together with the recognition of CD8 T cell responses against the human immunodeficiency virus (HIV-1). Since HIV-I infection modulates the expression of IFNs that the cellular environment, niche, there would render greater contribution of HLA-B leader sequences to generate HLA-E binding peptides to interact with the NKG2A receptor as well as interaction with the cognate immune system [48]. Based on this scenario, we believe that the presence of Met at P2 produces inhibition of killing favoring HIV progression whereas, Thr at P2 produces loss of inhibition of killing favoring activating receptors of NK cells that negatively influence HIV infection outcome.

Therefore it is important to emphasize that among the multiple genes that limit the progression to AIDS [4] HLA-E molecules loaded with either methionine of threonine in the second position of leader peptides should be studied not only in regard to their influence in infection outcome but also in the innate protection from HIV-1 infection. Furthermore, we will also use a SCID-hu model where a combination of TRIM-5a, CCR5, with threonine in the second position of the HLA-B leader peptide will protect transplants from infection. C) Cognate immunity. There is a large literature of this subject related to viremia progression and future investigations are needed in this respect to identify genes of protection from infection by HIV. These should include genetic markers involved in the role of CD4/CD8 in HIV-1 infection together with those genetic factors involved in mechanisms of innate immunity against HIV-1 viral infection or disease progression, such as those involving NK receptors interacting with their ligands. This mechanism also explains the slow progression to AIDS in a subgroup of patients that maintain normal CD4+ T cell counts but demonstrate viremia even after 15 years of infection [46].

In unpublished studies by Flores, et al [46] previous published results were reanalyzed [25] with 39 individuals of known immune status that included controllers of viremia with a short follow up (not longer than 2 years) The HLA-B alleles were grouped according the Bw4 or Bw6 public specificity encoding epitopes determined by residues 79-83 at the carboxyl-terminal end of the alpha 1 helix (Table 1). Some of the Bw6 alleles have threonine in the second position of their leader peptide, for example HLA-B*1801, B*4101, B*35 alleles, B*40 alleles, and B*15 alleles. These results were corroborated analyzing 98 HIV infected individuals with follow-up of viral loads and CD4+ T cell counts with progression or lack of progression to immunodeficiency. TT was associated with long-term progressors and with long-term non-progressors that maintain normal CD4 counts beyond 15 years from the date of infection.

Figure 1, summarizes possible variability of CD34 cells in the population, different age at time of thymic involution.

| CD34 (Stem cells) | Age at Thymus involution is variable.

Number of T cells in blood and lymphoid tissues is variable. | CD4/CD8 HIV infection
Variability due to Number of T cells, differentiation of stem cells, niche: ARGs
  a) Cognate Immunity
  b) Intrinsic Immunity
  c) NK cell receptors | a) Prevention of HIV viral entry
b) HIV infection variable outcome

1) Fast progressors
2) Intermediate progressors
3) Long-termn non progressors |

Figure 1. Diagram summarizes possible variability of CD34 cells in the population, different age at time of thymic involution.

This hypothesis can be studied as a cross-sectional study using the SCID mouse using peripheral blood and tissues available to investigate the number of CD4, CD8 and NK cells present at the time of short-term progression or long-term non-progression as a first step to study the role of NKG2A/ligand interaction in inhibition from killing, or actual killing, of HIV infected target cells as well as using the SCID-hu mouse model.

V. FUTURE STUDIES

Most studies of HIV infection outcome reported have described the importance of class I gene products particularly HLA-B. In this regard in a large number of HIV-1 infected individuals from southern Africa, it was reported that HLA-B alleles influence the potential co-evolution of HIV and HLA, providing advantage to pathogen defense mediated by $CD8^+$ T cells. These results were consistent with the findings in B-clade infected Caucasians with non-progression/low viral load or progression/high viral loads [48]. The authors claimed that the dominant effect for HLA-B alleles in HIV infection was unlikely due to differences in NK cell activation through the HLA-Bw4 motif. They mentioned the fact that it was reported that there is epistatic interaction between KIR3DS1 and some HLA-Bw4 alleles that mediated protection were entirely independent of this interaction [26]. Since there is a protective role of certain Bw4 alleles, B*2705, B*13 and B*44, as well as those that carry Ile-80, such a protective role should involve the mechanism described of NKG2A with the leader peptide threonine. See diagram. In both reports [26, 48], the authors did not consider the fact that there are two kinds of NK receptors, immunoglobulin receptors such as KIR3DS1 and the lectin receptors such as NKG2A. We have described herein that the alleles of Bw4, B*27 and B*57 as well as some Bw6 alleles carry threonine in the second position of their leader peptide that would interact with NKG2A. This mechanism should be studied together with the dominant role of HLA-B alleles influencing HIV specific CD8 T cell responses in the outcome of HIV infection. The use of cross-sectional studies has limitations of statistical power that are corrected when using longitudinal studies and this explains to some extent the major inconsistencies in published reports related to the HIV infection outcome. Perhaps the SCID mouse model could resolve some of these problems but this model also has limitations related to the fact that the experiments are limited to the particular tissues grafted and do not necessarily include the large number of ARGs present in individuals of the population at large [82, 83]. However, the engraftment of $CD34^+$ or even of $CD133^+$ cells give rise to multiple lineages of cells following spontaneous differentiation in vivo that may identify the source of ARGs.

ACKNOWLEDGEMENT

Supported by NIH grants HL29583 and HL59838, and also from the Dana Farber Cancer Institute. P.K. is a recipient of NIH grant HL079846.

REFERENCES

[1] Fauci, A.S. HIV and AIDS: 20 years of science. *Nat. Med,* 2003, 9, 839–843.
[2] O'Brien, S.J. and Moore, J. The effect of genetic variation in chemokines and their receptors on HIV transmission and progression to AIDS. *Immunol. Rev,* 2000, 177, 99–111.
[3] Carrington, M. and O'Brien, S.J. The influence of HLA genotype on AIDS. *Ann. Rev. Med,* 2003 54, 535–551.
[4] O'Brien SJ, Nelson GW. Human genes that limit *AIDS,* 2004, 36(6):565-74.
[5] Winkler C, An P, O'Brien SJ Patterns of ethnic diversity among the genes that influence AIDS. *Hum. Mol. Gene.,* 2004,13 Spec No 1:R9-19.
[6] Cao Y, Qin L, Zhang L, Safrit J, Ho DD.Virologic and immunologic characterization of long-term survivors of human immunodeficiency virus type 1 infection. *N. Engl. J. Med.,* 1995, 332(4):201-8.
[7] Pantaleo G, Menzo S, Vaccarezza M, Graziosi C, Cohen OJ, Demarest JF, Montefiori D, Orenstein JM, Fox C, Schrager LK, et al.Studies in subjects with long-term nonprogressive human immunodeficiency virus infection. *N. Engl. J. Med.,* 1995, 332(4):209-16.
[8] HRinaldo C, Huang XL, Fan ZF, Ding M, Beltz L, Logar A, Panicali D, Mazzara G, Liebmann J, Cottrill M, et al.igh levels of anti-human immunodeficiency virus type 1 (HIV-1) memory cytotoxic T-lymphocyte activity and low viral load are associated with lack of disease in HIV-1-infected long-term nonprogressors. *J. Virol.,* 1995, 69(9):5838-42.
[9] Harrer T, Harrer E, Kalams SA, Barbosa P, Trocha A, Johnson RP, Elbeik T, Feinberg MB, Buchbinder SP, Walker BD.Cytotoxic T lymphocytes in asymptomatic long-term nonprogressing HIV-1 infection. Breadth and specificity of the response and relation to in vivo viral quasispecies in a person with prolonged infection and low viral load. *J. Immunol.,* 1996, 156(7):2616-23.
[10] Kalams SA, Buchbinder SP, Rosenberg ES, Billingsley JM, Colbert DS, Jones NG, Shea AK, Trocha AK, Walker BD. Association between virus-specific cytotoxic T-lymphocyte and helper responses in human immunodeficiency virus type 1 infection, *J. Virol.,* 1999, 73(8):6715-20.
[11] Lisziewicz J, Rosenberg E, Lieberman J, Jessen H, Lopalco L, Siliciano R, Walker B, Lori F.Control of HIV despite the discontinuation of antiretroviral therapy. *N. Engl. J. Med.,* 1999, 340(21):1683-4.
[12] De Maria A, Ferraris A, Guastella M, Pilia S, Cantoni C, Polero L, Mingari MC, Bassetti D, Fauci AS, Moretta L. Expression of HLA class I-specific inhibitory natural killer cell receptors in HIV-specific cytolytic T lymphocytes: impairment of specific cytolytic functions, *Proc. Natl. Acad. Sci. USA,* 1997, 94(19):10285-8.

[13] C. Servet, L. Zitvogel and A. Hosmalin Dendritic Cells in Innate Immune Responses Against HIV Pp.739-756. *Current Molecular Medicine*, 2 (8), 2002.

[14] G. Herbein, A. Coaquette, D. Perez-Bercoff and G. Pancino Hosmalin. Macrophage Activation and HIV Infection: Can the Trojan Horse Turn into a Fortress? *Current Molecular Medicine*, 2002, 2 (8):723-738.

[15] Rappocciolo G, Birch J, Ellis SA.Down-regulation of MHC class I expression by equine herpesvirus-1. *J Gen Virol*, 2003, 84(2), 293-300.

[16] E.A. Trachtenberg and H.A. Erlich, A review of the role of the human leukocyte antigen (HLA) system as a host immunogenic factor influencing HIV transmission and progression to AIDS. In: B.T.K. Korber, C. Brander, B.F. Haynes, J.P. Moore, R.A. Koup, C. Kuiken, B.D. Walker and D.I. Watkins, Editors, HIV Molecular Immunology 2001, Theoretical Biology and Biophysics Group (2001), pp. 143-160.

[17] Delgado JC, Leung JY, Baena A, Clavijo OP, Vittinghoff E, Buchbinder S, Wolinsky S, Addo M, Walker BD, Yunis EJ, Goldfeld AE. The -1030/-862-linked TNF promoter single-nucleotide polymorphisms are associated with the inability to control HIV-1 viremia. *Immunogenetics*, 2003, 55(7):497-501.

[18] Folks TM, Clouse KA, Justement J, Rabson A, Duh E, Kehrl JH, Fauci AS.Tumor necrosis factor alpha induces expression of human immunodeficiency virus in a chronically infected T-cell clone. *Proc. Natl. Acad. Sci. USA*, 1989, 86(7):2365-8.

[19] Marshall WL, Brinkman BM, Ambrose CM, Pesavento PA, Uglialoro AM, Teng E, Finberg RW, Browning JL, Goldfeld AE. Signaling through the lymphotoxin-beta receptor stimulates HIV-1 replication alone and in cooperation with soluble or membrane-bound TNF-alpha. *J. Immunol.*, 1999, 162(10):6016-23.

[20] Poli G, Kinter A, Justement JS, Kehrl JH, Bressler P, Stanley S, Fauci AS.Tumor necrosis factor alpha functions in an autocrine manner in the induction of human immunodeficiency virus expression. *Proc. Natl. Acad. Sci. USA*, 1990, 87(2):782-5.

[21] Gao, X. et al. Effect of a single amino acid change in MHC class I molecules on the rate of progression to AIDS. *N. Engl. J. Med*, 2001, 344, 1668-1675.

[22] O'Brien, S.J., Gao, X. and Carrington, M. HLA and AIDS: A cautionary tale. *Trends Mol. Med.*, 2002, 7, 379-381.

[23] Carrington, M. et al. HLA and HIV-1: Heterozygote advantage and B*35-Cw*04 disadvantage. *Science*, 1999, 283, 1748-1752.

[24] Tang, J.M. et al. HLA class I homozygosity accelerates disease progression in human immunodeficiency virus type I infection. *AIDS Res. Hum.*, 1999, Retroviruses 15, 317-324.

[25] Flores-Villanueva, P.O. et al. Control of HIV-1 viremia and protection from AIDS are associated with HLA-Bw4 homozygosity. *Proc. Natl. Acad. Sci. USA* 98, 2001, 5140-5145.

[26] Martin MP, Gao X, Lee JH, Nelson GW, Detels R, Goedert JJ, Buchbinder S, Hoots K, Vlahov D, Trowsdale J, Wilson M, O'Brien SJ, Carrington M.Epistatic interaction between KIR3DS1 and HLA-B delays the progression to AIDS. *Nat. Genet*, 2002, 4:429-34.

[27] Qing M, Li T, Han Y, Qiu Z, Jiao Y.Accelerating effect of human leukocyte antigen-Bw6 homozygosity on disease progression in Chinese HIV-1-infected patients. *J. Acquir. Immune Defic. Syndr*, 2006, 41(2):137-9.

[28] Kaslow RA, Tand J, Dorak MT, Tang s, Musonda R, Karita E, Wilson C, allen S. Homozigosity for HLA-Bw4 is not associated with protection of HIV-1 infected persons in African ancestry. *Conference retroviruses opportunistic infect*, 2002, Feb 24-28;9: abstract no. 320-W.

[29] Best, S., Le Tissier, P., Towers, G., and Stoye, J. P. (1996). Positional cloning of the mouse retrovirus restriction gene Fv1. Nature 382, 826-829.

[30] Sheehy, A. M., Gaddis, N. C., Choi, J. D., and Malim, M. H. (2002). Isolation of a human gene that inhibits HIV-1 infection and is suppressed by the viral Vif protein. Nature 418, 646-650.

[31] Stremlau, M., Owens, C. M., Perron, M. J., Kiessling, M., Autissier, P., and Sodroski, J. (2004). The cytoplasmic body component TRIM5alpha restricts HIV-1 infection in Old World monkeys. Nature 427, 848-853.

[32] Li, Y., Li, X., Stremlau, M., Lee, M., and Sodroski, J. (2006). Removal of arginine 332 allows human TRIM5alpha to bind human immunodeficiency virus capsids and to restrict infection. J Virol 80, 6738-6744.

[33] Yap, M. W., Nisole, S., and Stoye, J. P. (2005). A single amino acid change in the SPRY domain of human Trim5alpha leads to HIV-1 restriction. Curr Biol 15, 73-78.

[34] Javanbakht H, An P, Gold B, Petersen DC, O'Huigin C, Nelson GW, O'Brien SJ, Kirk GD, Detels R, Buchbinder S, Donfield S, Shulenin S, Song B, Perron MJ, Stremlau M, Sodroski J, Dean M, Winkler C. Effects of human TRIM5alpha polymorphisms on antiretroviral function and susceptibility to human immunodeficiency virus infection, *Virology*, 2006, 354(1):15-27.

[35] Winkler, C.A. and O'Brien, S.J. *AIDS restriction genes in human ethnic groups: An assessment. in AIDS in Africa* (eds. Essex, M., Mboup, S., Kanki, P.J., Marlink, R. and Tlou, S.D.) 2nd edn. Kluwer Academic, New York, 2002.

[36] Speelmon EC, Livingston-Rosanoff D, Li SS, Vu Q, Bui J, Geraghty DE, Zhao LP, McElrath MJ.Genetic association of the antiviral restriction factor TRIM5alpha with human immunodeficiency virus type 1 infection. *J. Virol.,* 2006, 80(5):2463-71.

[37] Scott-Algara D, Truong LX, Versmisse P, David A, Luong TT, Nguyen NV, Theodorou I, Barre-Sinoussi F, Pancino G.Cutting edge: increased NK cell activity in HIV-1-exposed but uninfected Vietnamese intravascular drug users. *J. Immunol.,* 2003, 171(11):5663-7.

[38] Collins KL, Chen BK, Kalams SA, Walker BD, Baltimore D.HIV-1 Nef protein protects infected primary cells against killing by cytotoxic T lymphocytes. *Nature,* 1998, 391(6665):397-401.

[39] McCutcheon JA, Gumperz J, Smith KD, Lutz CT, Parham P.Low HLA-C expression at cell surfaces correlates with increased turnover of heavy chain mRNA. *J. Exp. Med.,* 1995,181(6):2085-95.

[40] Cohen GB, Gandhi RT, Davis DM, Mandelboim O, Chen BK, Strominger JL, Baltimore D. The selective downregulation of class I major histocompatibility complex proteins by HIV-1 protects HIV-infected cells from NK cells. *Immunity,* 1999,10(6):661-71.

[41] Liu K, Kao KJ. Mechanisms for genetically predetermined differential quantitative expression of HLA-A and -B antigens. *Hum. Immunol.* 2000 Aug; 61(8):799-807.

[42] Katsuyama Y, Ota M, Ando H, Saito S, Mizuki N, Kera J, Bahram S, Nose Y, Inoko H.Sequencing based typing for genetic polymorphisms in exons, 2, 3 and 4 of the MICA gene. *Tissue Antigens,* 1999, 54(2):178-84.

[43] Natarajan K, Dimasi N, Wang J, Mariuzza RA, Margulies DH.Structure and function of natural killer cell receptors: multiple molecular solutions to self, nonself discrimination. *Annu. Rev. Immunol.,* 2002, 20:853-85.

[44] Phillips JH, Gumperz JE, Parham P, Lanier LL. Superantigen-dependent, cell-mediated cytotoxicity inhibited by MHC class I receptors on T lymphocytes. *Science,* 1995, 268(5209):403-5.

[45] Borrego F, Ulbrecht M, Weiss EH, Coligan JE, Brooks AG.Recognition of human histocompatibility leukocyte antigen (HLA)-E complexed with HLA class I signal sequence-derived peptides by CD94/NKG2 confers protection from natural killer cell-mediated lysis. *J. Exp. Med.,* 1998, 187(5):813-8.

[46] Flores-Villanueva, Yunis E, Buchbinder S, Vittinghoff E, walker B. Association of two copies of HLA-B alleles encoding HLA-E binding peptides with Threonine at postion 2 with control of viremia and progression to AIDS. Unpublished.

[47] Brooks AG, Borrego F, Posch PE, Patamawenu A, Scorzelli CJ, Ulbrecht M, Weiss EH, Coligan JE.Specific recognition of HLA-E, but not classical, HLA class I molecules by soluble CD94/NKG2A and NK cells. *J. Immunol.,* 1999, 162(1):305-13.

[48] Kiepiela P, Leslie AJ, Honeyborne I, Ramduth D, Thobakgale C, Chetty S, Rathnavalu P, Moore C, Pfafferott KJ, Hilton L, Zimbwa P, Moore S, Allen T, Brander C, Addo MM, Altfeld M, James I, Mallal S, Bunce M, Barber LD, Szinger J, Day C, Klenerman P, Mullins J, Korber B, Coovadia HM, Walker BD, Goulder PJ. Dominant influence of HLA-B in mediating the potential co-evolution of HIV and HLA. *Nature,* 2004, 432(7018):769-75.

[49] Lanier LL.Natural killer cells: from no receptors to too many. *Immunity,* 1997, (4):371-8.

[50] Miller JD, Weber DA, Ibegbu C, Pohl J, Altman JD, Jensen PE. Analysis of HLA-E peptide-binding specificity and contact residues in bound peptide required for recognition by CD94/NKG2. *J. Immunol.*, 2003, 171(3):1369-75.

[51] WWW.anthonynolan.com/HIGseq/pep).

[52] Koka, PS; Reddy, ST. Cytopenias in HIV infection: Mechanisms and alleviation of hematopoietic inhibition. *Curr. HIV Res.*, 2004, 2, 275-282.

[53] Miles, SA; Mitsuyasu, RT; Moreno, J; Baldwin, G; Alton, NK; Souza, L; Glaspy, JA. Combined therapy with recombinant granulocyte colony-stimulating factor and erythropoietin decreases hematologic toxicity from zidovudine. *Blood*, 1991, 77, 2109-2117.

[54] Miles, SA; Lee, S; Hutlin, L; Zsebo, KM; Mitsuyasu, RT. Potential use of Human stem cell factor as adjunctive therapy for Human immunodeficiency virus-related cytopenias. *Blood*, 1991, 78, 3200-3208.

[55] Ratner, L. Human immunodeficiency virus-associated autoimmune thrombocytopenic purpura: A review. *Am. J. Med.*, 1989, 86, 194-198.

[56] Fauci, AS. Host factors and the pathogenesis of HIV-induced disease. *Nature*, 1996, 384, 529-534.

[57] Harbol, AW; Liesveld, JL; Simpson-Haidaris, PJ; Abboud, CN. Mechanisms of cytopenia in human immunodeficiency virus infection. *Blood Rev.*, 1994, 8, 241-251.

[58] Shen, H; Cheng, T; Preffer, FI; Dombkowski, D; Tomasson, MH; Golan, DE; Yang, O; Hofmann, W; Sodroski, JG; Luster, AD; Scadden, DT. Intrinsic Human immunodeficiency virus type 1 resistance of hematopoietic stem cells despite coreceptor expression. *J. Virol.*, 1999, 73, 728-737.

[59] Koka, PS; Jamieson, BD; Brooks, DG; Zack, JA. Human immunodeficiency virus type-1 induced hematopoietic inhibition is independent of productive infection of progenitor cells in vivo. *J. Virol.* 1999, 73, 9089-9097.

[60] Koka, PS; Fraser, JK; Bryson, Y; Bristol, GC; Aldrovandi, GM; Daar, ES; Zack, JA. Human immunodeficiency virus type 1 inhibits multilineage hematopoiesis in vivo. *J. Virol.*, 1998, 72, 5121-5127.

[61] Jenkins, M; Hanley, MB; Moreno, MB; Wieder, E; McCune, JM. Human immunodeficiency virus-1 infection interrupts thymopoiesis and multilineage hematopoiesis in vivo. *Blood*, 1998, 91, 2672-2678.

[62] Koka, PS; Kitchen, CM; Reddy, ST. Targeting c-Mpl for revival of human immunodeficiency virus type 1-induced hematopoietic inhibition when CD34+ progenitor cells are re-engrafted into a fresh stromal microenvironment in vivo. *J. Virol.*, 2004, 78, 11385-11392.

[63] Sankaran S, Guadalupe M, Reay E, George MD, Flamm J, Prindiville T, Dandekar S. Gut mucosal T cell responses and gene expression correlate with protection against disease in long-term HIV-1-infected nonprogressors. *Proc. Natl. Acad. Sci. USA.*. 2005, 102(28):9860-5.

[64] Stier S, Cheng T, Dombkowski D, Carlesso N, Scadden DT. 2002. Notch1 activation increases hematopoietic stem cell self-renewal in vivo and favors lymphoid over myeloid lineage outcome. *Blood* 99: 2369-2378.

[65] Pestina TI, Jackson CW. 2003. Differential role of Stat5 isoforms in effecting hematopoietic recovery induced by Mpl-ligand in lethally myelosuppressed mice. *Exp. Hematol.* 31: 1198-1205.

[66] Schulze H, Ballmaier M, Welte K, Germeshausen M. 2000. Thrombopoietin induces the generation of distinct Stat1, Stat3, Stat5a and Stat5b homo- and heterodimeric complexes with different kinetics in human platelets. *Exp. Hematol.* 28: 294-304.

[67] Zeng H, Masuko M, Jin L, Neff T, Otto KG, Blau CA. 2001. Receptor specificity in the self-renewal and differentiation of primary multipotential hematopoietic cells. *Blood* 98: 328-334.

[68] Bradley HL, Couldrey C, Bunting KD. 2004. Hematopoietic-repopulating defects from STAT5-deficient bone marrow are not fully accounted for by loss of thrombopoietin responsiveness. *Blood* 103: 2965-2972.

[69] Goncalves F, Lacout C, Villeval JL, Wendling F, Vainchenker W, Dumenil D. 1997. Thrombopoietin does not induce lineage-restricted commitment of Mpl-R expressing pluripotent progenitors but permits their complete erythroid and megakaryocytic differentiation. *Blood* 89:3544-3553.

[70] Kaushansky K. 1998. Thrombopoietin and the hematopoietic stem cell. *Blood* 92:1-3.

[71] Kaushansky K, Lin N, Grossman A, Humes J, Sprugel KH, Broudy VC. 1996. Thrombopoietin expands erythroid, granulocyte-macrophage, and megakaryocytic progenitor cells in normal and myelosuppressed mice. *Exp. Hematol.* 24:265-269.

[72] Silvestris F, Cafforio P, Tucci M, Dammacco F. 2002. Negative regulation of erythroblast maturation by Fas-L+/TRAIL+ highly malignant plasma cells: a major pathogenic mechanism of anemia in multiple myeloma. *Blood* 99:1305-1313.

[73] Solar GP, Kerr WG, Zeigler FC, Hess D, Donahue C, de Sauvage FJ, Eaton DL. 1998. Role of c-mpl in early hematopoiesis. *Blood* 92:4-10.

[74] Schuringa JJ, Chung KY, Morrone G, Moore MAS. 2004. Constitutive activation of STAT5A promotes human hematopoietic stem cell self-renewal and erythroid differentiation. *J. Exp. Med.* 200: 623-635.

[75] Mingari MC, Moretta A, Moretta L. Regulation of KIR expression in human T cells: a safety mechanism that may impair protective T-cell responses. *Immunol. Today.* 1998, 19(4):153-7.

[76] Ponte M, Bertone S, Vitale C, Tradori-Cappai A, Bellomo R, Castriconi R, Moretta L, Mingari MC. Cytokine-induced expression of killer inhibitory receptors in human T lymphocytes. *Eur. Cytokine Netw.* 1998,9(3 Suppl):69-72.

[77] Mingari MC, Ponte M, Bertone S, Schiavetti F, Vitale C, Bellomo R, Moretta A, Moretta L. HLA class I-specific inhibitory receptors in human T lymphocytes: interleukin 15-induced expression of CD94/NKG2A in superantigen- or alloantigen-activated CD8+ T cells. *Proc. Natl. Acad. Sci. USA.* 1998 Feb 3;95(3):1172-7.

[78] Fodil N, Pellet P, Laloux L, Hauptmann G, Theodorou I, Bahram S. MICA haplotypic diversity. *Immunogenetics.* 1999, 49(6):557-60.

[79] Lee N, Llano M, Carretero M, Ishitani A, Navarro F, Lopez-Botet M, Geraghty DE. HLA-E is a major ligand for the natural killer inhibitory receptor CD94/NKG2A. *Proc. Natl. Acad. Sci. USA.* 1998, 95(9):5199-204.

[80] Llano M, Lee N, Navarro F, Garcia P, Albar JP, Geraghty DE, Lopez-Botet M. HLA-E-bound peptides influence recognition by inhibitory and triggering CD94/NKG2 receptors: preferential response to an HLA-G-derived nonamer. *Eur. J. Immunol.* 1998, 28(9):2854-63.

[81] Andre P, Brunet C, Guia S, Gallais H, Sampol J, Vivier E, Dignat-George F. Differential regulation of killer cell Ig-like receptors and CD94 lectin-like dimers on NK and T lymphocytes from HIV-1-infected individuals. *Eur. J. Immunol.* 1999, 29(4):1076-85.

[82] Sundell IB, Koka PS. Chimeric SCID-hu model as a human hematopoietic stem cell host that recapitulates the effects of HIV-1 on bone marrow progenitors in infected patients. *J. Stem Cells* 2006; 1(4): 283-300.

[83] Melkus MW, Estes JD, Padgett-Thomas A, Gatlin J, Denton PW, Othieno FA, Wege AK, Haase AT, Garcia JV. Humanized mice mount specific adaptive and innate immune responses to EBV and TSST-1. *Nature Med.* 2006 12(11): 1316-1322.

Index

C

F

Index

J

N

O

P

Q

R

S

T